CREATING VALUE BY DESIGN
# THOUGHTS
by Stefano Marzano

Creating Value by Design

**Thoughts** by Stefano Marzano

©1998 Royal Philips Electronics

Published by V+K Publishing,

Blaricum, The Netherlands


Published by Lund Humphries Publishers

Park House, 1 Russell Gardens

London NW11 9NN

British Library Cataloguing in Publication Data

A catalogue record for this book is available from the British Library


ISBN 90 6611 731 1 (V+K Publishing)

ISBN 0 85331 757 7 (Lund Humphries Publishers)

Nugi 926


Distributed in the USA by

Antique Collectors' Club

Market Street Industrial Park

Wappingers Falls

NY 12590

USA

A more detailed pictorial impression of the products, services, communications and projects developed by the Philips Design Group within the context of High Design and on the basis of ideas described here can be found in the companion volume, *Creating Value by Design. Facts.*

# Contents

# Preface

Every morning, the people of the Balinese fishing village where I'm staying place a small offering of fruit in a basket of banana leaves in the fork of a tree – an ancient gesture to appease the gods. How natural, I thought; how reassuring that traditional values are being maintained in the face of advancing Western culture and globalisation. That is, until one morning I saw, amidst the fruit, a small, familiar blue-and-white wrapped package. On closer examination, it turned out to be (ironically, both appropriately and inappropriately) that very Western icon of the swaying-palms-and-silver-beaches desert island… a coconut-filled chocolate bar. Few of us in the industrialised society which gave birth to such chocolate bars would ever think of imbuing them with a sacramental value like this, although we regularly give chocolate as a gift. Yet here, on Bali, a new, global product had been smoothly and silently incorporated into an ancient tradition.

Clearly, up to a point, human values are relative. But are there any values that are universal? And if so, what are they? These are ancient questions, but they are questions that are also becoming increasingly urgent, as globalisation affects the lives of more and more people. The reflections in this book – fragmented thoughts on related themes, rather than any coherent and worked-out theory – represent my own struggle to come to grips with the urgent issues of sustainability facing society in the coming decades, and to formulate practical strategies for dealing with them.

**Design in the Third Wave**  Whichever hemisphere we live in, we are living at a time of transition. I do not mean the merely accidental transition from one millennium into another, but the much more significant transition from what Alvin Toffler called the 'second wave' to the 'third wave' – from the Industrial Age to the Information Age. The revolution that is taking place around us (some call it the digital revolution) will cause as much upheaval in the way we live and work as the revolution that, only two centuries ago, took our ancestors from the Agricultural Age (Toffler's first wave) into the Industrial Age.

Society, consumers and products are all changing radically – and with them the nature and scope of design. We are moving from local orientation to global orientation, from predictable to unpredictable consumer behaviour, and from highly tangible, even cumbersome products to those that are tiny and barely more than packaged information. The advent of digitalisation affects the design of all products and services, whether they are themselves digital or not. The growing complexity of the world means that people are calling out for overall simplification, wherever it can be achieved. The speed generated by digital technologies is becoming the normal pace for activities of any sort. And while the globalisation enabled by new digital technologies is leading to a deeper appreciation and enjoyment of cultural diversity, it is not doing away with the need for products to be seen as having local significance.

The challenge for designers, and indeed for everyone in societies that are entering the Third Wave, is to discover the new relevant benefits and qualities – the qualities that products and services will need to have if they are to fulfil the aspirations and dreams of those who use them.

**Protagonists in designing the future** Finding these qualities is a shared responsibility: all those with power need to use that power wisely and responsibly to this end, where they can. Designers, through their designs, exert a powerful influence on society. If we take this ethical responsibility seriously, we will want to be active as protagonists in shaping the future, making sure that the results of our work advance people's personal growth, create a healthy living environment, both natural and artificial, and engender a relation between people and their environment that is maximally harmonious.

**Goals, strategies** In this anthology, which brings together thoughts developed and written down over a period of six or more years, I explore what I see (at this time, at least) as the major issues facing design in the Third Wave.

My aim has been, in the first place, to clarify for myself and those I work with where the urgent problems lie and what we can do to tackle them. My method has been to immerse myself in the issues, to reflect on them, analyse them, and wrestle with them, viewing them from all sides, in an act of philosophical exploration. I have tried not to be content with taking the problem as a given, but have always tried to question the question, as it were, allowing my train of thought to take me in often unexpected directions. The image that comes into my mind is that of someone feeling their way around in a dense fog. From time to time, you can discern outlines, feel shapes, see the welcome glow of a lamp. You have a sense of where you are, and that taking this turn or that will (surely?) bring you closer to your destination. At other times, you stumble blindly, hands stretched out in front of you, and the most you see is a sign that looks disconcertingly like one you have seen before. Are we back where we started? Or does the fact that we are approaching from a different direction make its meaning clearer now? This book, then, is a report on work in progress, of an unfinished – indeed, unfinishable – journey.

In the second place, I hope through these reflections to encourage others (my colleagues, the design community, our industrial and commercial clients, and the end-users of our products) to think along with me. For it is clear that no single individual, or small group, can hope to make any significant impact alone.

In the first piece, 'Flying over Las Vegas', I outline the problems which characterise the current 'crisis' situation, such as information overkill, consumer fragmentation and confusion, 'transversal' behaviour, pressure on the environment, and explore how they have come about. I then sketch the elements of a programme of action and introduce the concept of 'High Design', a multi-skilled, holistic approach to design which, by incorporating insights from the human sciences and anthropology, enables designers to accept the challenge of creating products which play a full part in a harmonious whole of human beings in their natural and artificial environment – products, in other words, that are relevant in a truly comprehensive sense. The High Design process is a practical tool. But, like any tool, it is only helpful if you know what you want to make with it. How do we discover what qualities people will feel are relevant to them? How do we find out what products they will welcome?

In 'The Smile in My Grandfather's Mirror', I discuss ways of addressing this very practical problem, and describe the strategy that we at Philips Design have adopted – and are continually refining. In the next eight pieces, I look at issues in a number of aspects of life: the home, healthcare, light, the natural urban environment, mobility and port-

ability. I've grouped the issues around topics relevant to the company I work for, but I believe the underlying principles are much more broadly applicable. No particular significance should be attached to the order in which they appear.

**Home**  The home is the most intimately human domain. In the essay entitled 'Into the Era of Soul', I consider how the 'soul' of television – its meaningful elements – can satisfy a vast range of needs of individuals within the new polycentric domestic setting, taking over many functions formerly fulfilled by other objects and circumstances, which, for one reason or another, we have banished from our lives. This essay formed part of the thinking that lay behind our project called *Television at the Crossroads* (1994). Many people find new technologies daunting and an ugly intrusion in the home. The question of how they can be seamlessly integrated into the domestic environment without disrupting traditional values is the issue I tackle in 'Designing Reconciliation'. We explored it in practical detail in the associated project, *New Objects, New Media, Old Walls* (1995). This project resulted later in the striking Plugged Furniture range, produced commercially in collaboration with Leolux. Then, in 'A Design Recipe for These Times', I tell the background to the Philips-Alessi line of kitchen appliances, another commercial success resulting from a similar project.

**Healthcare and well-being**  The holistic approach to health – *mens sana in corpore sano* – goes back to way beyond the Romans. But during the present century, as technology has assumed an increasingly central position in healthcare, the effect of the psyche on bodily illness has tended to be pushed aside. Now, there's a general realisation that, although technologically healthcare is better than it's ever been, we've been less effective than our forebears in providing reassurance and encouragement to those who are ill. Designers are clearly not physicians, but is there any way we can make a better contribution to patients' sense of well-being? Do we take sufficient account of the effect the design of medical equipment can have on the patients' peace of mind? In 'Humanware in Healthcare', I consider how the interests of the various parties involved in the purchase, use, maintenance and experience of medical equipment can be met through design. The importance of preventive medicine is also becoming increasingly appreciated, both by the medical profession and the general public. This includes paying greater attention to matters of diet, personal hygiene and stress reduction. Individuals are realising that their future health and sense of well-being lie to a considerable extent in their own hands. The sense of taking personal charge of one's own condition is extending beyond the traditional boundaries of medicine. As a holistic approach to health – or 'wellness' – becomes the norm, new personal care products will allow people to feel good about themselves both internally and externally, making care for 'body and soul' another area in which designers and manufacturers can contribute significantly to enhancing the quality of life.

**Light**  Light has an incredibly strong psychological and emotional effect on people; but for a manufacturer of lighting equipment, such as Philips, it is perhaps all too easy to concentrate on technical matters, leaving the use of light to create emotional effects to interior designers, stage lighting designers and architects. In 'New Frontiers in Lighting Design', I look at factors that designers of lighting equipment could perhaps take into account

to make their creations more 'human-focused'. And in 'City Treasures', I consider the significant role that lighting can play in enhancing the urban environment.

**Natural urban environment**  It takes more to create a 'natural' or human-focused urban environment than sophisticated flood-lighting, of course. As part of a proposal for the redevelopment of an industrial historical building, I discuss, in 'Harmoniopolis', the sort of urban environment that would provide people with a quasi-urban experience in which the 'natural' or intuitive elements are maximised while still maintaining the benefits of a complex society.

**On the Move**  Miniaturisation will increasingly result in many technologies being incorporated if not into the body itself then at least into our 'second skin', our clothing. This will make us even more independent of place, and will allow us to lead a much fuller life on the move. In 'Uniformity, Diversity and the New Nomads', I place this phenomenon in its evolutionary context and consider its implications.

**Ethics and Commercial Reality**  The programme of action that, in essence, I put forward in the course of these reflections can only succeed and be true to itself if it is pursued within an ethical framework and in the context of commercial realities. In several other pieces, therefore, I look at the sorts of ethical and commercial considerations that need to be taken into account in the Third Wave.

In 'Chocolate for Breakfast', I challenge the design community to take an ethical stance on the environment and the question of sustainable development: designers should not merely pay lip service to the issues, but should join forces with the media to persuade governments and other authorities to create the right infrastructure and other conditions for sustainable development. Then, 'In Search of a Sustainable Society' looks at the particular role that information technology can play in the attainment of sustainability. I ponder the possibility that a coincidence of historical timing – the simultaneous rise of information technology and the development of an environmental crisis in the late twentieth century – is opening up the hopeful prospect of a 'New Modernity', a society in which progress is still possible, but where progress is seen as a question of improving quality rather than of expanding quantity. In 'The Metamorphosis of Products', I describe four strategies for developing charismatic products that are compatible with sustainability. These ideas were worked out in some detail in a project which formed part of a larger international project under the name *The Solid Side.* Machines are becoming increasingly interactive and also emancipated from their human creators. This imposes a heavy responsibility on designers to make sure that tomorrow's products 'behave' as we would wish them to do. In other words, we need ensure we create 'well-brought-up machines'. This issue is addressed in 'Talk to me, Moses!'

Idealism without practical application is worthless. The system within which we live requires of industrial designers that they ensure that their products are commercially successful. The companies for whom they work exist to provide investors with a good return on their investment. An ailing company is unable to help anyone in society: it is a burden to all. The High Design programme is therefore not a mere philosophical exercise; it is a very down-to-earth approach to design. It allows us to create products that, by meeting certain conditions, have an excellent chance of com-

mercial success. Such products should help people accomplish and experience the things they want to accomplish and experience; they should allow them to do these things with minimum effort and in natural, intuitive ways; and, in their manufacture, operation and effects, such products should be fully in line with their users' fundamental value systems.

But designing products that sell well is not the only way designers can serve both their companies and the public. In today's crowded market place, a strong brand is a boon both to the company and the customer. Shoppers, overwhelmed by the range of choice, thankfully reach for the familiar brand. The manufacturer, in turn, is grateful to stand out in the crowd. In the final piece, 'The Monk and the Machine Gun', I reflect on the nature of branding and the role played by design in helping companies give tangible expression to the intangible qualities that constitute their distinctive offering to the consumer.

Finally, for discussion and feedback, I'd like to express my thanks to colleagues both past and present (particularly, of course, at Philips), to audiences at conferences and symposiums at which I presented earlier versions of my ideas, and to family, friends and acquaintances, who were patient enough to take my ideas seriously, even on light-hearted occasions.

*Stefano Marzano*
Jembaran (Bali), January 1998

# Flying over Las Vegas

Some months ago, I was on board an aeroplane. Looking out of the window at the ground below, I could see the bright neon lights of a city standing in the middle of a desert. As you may already have guessed, its name was Las Vegas. After the plane landed, I found myself in an artificial environment whose inhabitants appeared to be motivated by greed. In Las Vegas, the gambling casinos are specifically designed to encourage this: daylight is entirely shut out and there are no clocks on the walls. All activity is geared to the lust for money, a lust which is seldom satisfied. Money and materialism are the city's guiding principles; and this probably explains why most of the people who go there to play the slot machines seem to be so unhappy.

On my visit, I was confronted with an awesome vision of a future hell, a place or state of mind which recalled the writer Henry Miller's expression, "the air-conditioned nightmare". Las Vegas seems to give us a warning of what the future could be like, if we're not careful: the type of world to which environmental carelessness and materialism supported by technology could lead us. But it's not technology that determines the destiny of humanity, but rather people themselves, in how they decide to use this technology. The future doesn't just happen by itself. It can be influenced by those who are prepared to shoulder the responsibility of making decisions today.

That means that we, too, can participate in the shaping of this future. By virtue of the enormous number of products they put onto the market, large companies play a major role in determining the quality of our lives. Such corporations should therefore acknowledge their responsibility and become conscious of their power. Those of us who work for them must accept our share of that responsibility.

**The new order**  Many different challenges are confronting the human race. No-one can deny that the last decade has seen dramatic changes in political, social, economical and environmental terms; and we're all naturally somewhat apprehensive about the future. And yet there is also hope and optimism. The fall of the Berlin Wall in 1989 was more than just a political event. It marked the end of an era in which the world had been deliberately – and skilfully – divided into two. As the philosopher Herbert Marcuse (who died some ten years before the fall of the Wall) remarked in his book *One-Dimensional Man*, people in the western democracies were brought up to fear Communist aggression: similarly, people in Communist countries were taught to live in fear of capitalist aggression. This artificial division into Good and Evil undoubtedly served to keep nations united. Now that this division no longer exists, however, our community loyalties are shifting. The world is fragmented. We're discovering new local, ethnic and regional groupings; and this helps to explain the current crises of national identity.

The old system blinded us to the reality of the situation. It prevented us from realising that, instead of belonging to either the East or the West, we're in fact members of a global community. We were prevented from seeing that the problems of this community must be solved on a global basis. The problem is, however, that we are still far from being a single global community. Countries are still divided into two basic types: on the one hand, the Industrial Triad of North America, Japan and Europe (together with one or two additions, such as Australia and New Zealand), and on the other, all the other countries of the world.

As we approach the millennium, we're seeing the effects of a gross imbalance in material circumstances around the world. The idyll of the Triad countries is beginning to crumble, not only because we're being challenged by our fellow human beings, but also because our physical, natural environment is showing severe signs of strain.

**Saturation point**    The Triad countries have reached saturation point. We're constantly being bombarded by messages – from advertising and the news media – that slowly but surely are ceasing to have any meaning. Every day the mass media invade our thoughts with a constant flow of confusing images. These may appear attractive, but in reality they simply cover a void. We're confronted by a Malthusian glut of information, and the result is semantic and semiotic pollution.

A similar glut of information is apparent in the products we use. They confuse us with their vast number of functions and, again, their seductive exterior merely covers a void. It is this void, or silence, which prevents us from entering into any meaningful relationship with the product: we're unable to 'feel' the object, just as we're unable to cope with the mass media images. In both cases, the effect is one of mental pollution; and as with environmental pollution, the fundamental cause is an undue emphasis on quantity at the expense of quality.

Parallel to the fragmentation of community loyalties we're witnessing around the world, we're also seeing a segmentation of consumer loyalty. Today's consumers are increasingly individualistic. Whereas they once fell into clearly defined national, cultural or social groups (e.g., the yuppies) and acquired products (often status symbols) in order to affirm their membership of one of these groups, today they're 'transversal' in their behaviour. In the words of Francesco Morace, of the Domus Academy in Milan, today's consumers are "complex, flexible and multi-dimensional". They'll cross the social group boundaries several times during the course of the day. A highly rational purchase will be followed by a piece of impulse buying. They'll eat in a luxurious restaurant one day and in a cheap pizzeria the next; wear an expensive Rolex watch along with jeans and sneakers. The consumer is moving away from group style towards individual style. Consumers have become individuals, and although they exhibit certain similarities, they have become less predictable.

**The unlimited market**    In both the political and the consumer scenario, chaos is the new order. And yet, there is no reason for pessimism. If our limited market has now reached saturation point, and the metaphorical wall has been knocked down, then the next logical step is the unlimited market. And this should be an exciting place to be.

In the past, the emphasis was on quantity: products were impressive for their number of functions, the number of gears or the number of programmes. The car was judged on its speed, the stereo system for its volume, and so on. Perhaps because of this quantitative, materialistic approach, we have now reached the limits to growth. Concentration on quantity is therefore no longer possible. Instead, we must shift to quality. We must focus on the 'soft' values – the aspects which will make the consumer experience richer and more meaningful, the values which will promote humanity's cultural growth.

We should therefore make sure we create *relevant* objects – products which use technology not for technology's sake, but to encourage the individual's cultural growth; products which promote the amplification of the senses and the individual's capabilities. And I believe the rapid advances being made in design technology are such that this is genuinely possible.

**Re-interiorisation**  If we consider the history of humanity's progress, we see that the evolution of homo sapiens has been accompanied by a constant process of exteriorisation. The limits to our physical strength prompted us to start using tools. With the passage of time, these became larger and more complex. The development of transport was also part of this exteriorisation process. First, horses replaced our legs; then came the wheel; and finally, various types of engines have resulted in the train, the car and the aeroplane. In the twentieth century, this exteriorisation process has been applied not only to our physical powers, but also to our mental faculties: from the simple adding machine we've developed the computer. We've even exteriorised our reproductive capacities. The idea of the test-tube baby, which was mocked in Aldous Huxley's futuristic novel *Brave New World* (1932), is now an everyday reality, and human cloning may not be far behind.

There are basically two types of exteriorisation: a 'heavy' variety, represented by mechanical machines, and a 'light' one, represented by electronic and digital technology. In recent decades, we've witnessed the increasing miniaturisation of products: the portable computer, the pocket-sized television, mini stereo-systems, and so on. This development conjures up a number of fascinating scenarios for the future. One of these is that we will embark on a phase of re-interiorisation. The Walkman, for example, is, in a sense, a pocket-sized body prosthesis, marking the return to 'skin level' of a previous result of the exteriorisation process. If miniaturisation continues, then products will occupy less and less space in the home. They may (indeed, probably will) merge invisibly into the domestic environment, leaving room for more meaningful and relevant objects, objects that represent memories, actualities and culture, making way for personal effects, chairs or paintings that perhaps belonged to a relative or ancestor.

**Cultural totems**  Not all new products will need to be so self-effacing, however. They may become cultural totems in themselves. Some examples of this were shown at the *Ambiente Electronica* show in Berlin (1991), where Philips asked a number of eminent designers to create a symbiosis between intelligent technology and progressive trends in interior design under the motto 'Design and materials in harmony with their environment'.

In one scenario, the television was fully integrated into the bedroom setting. The Parisian designers, Elizabeth Garouste and Mattia Bonetti, envisaged the bedroom as a theatre of dreams. A place for sleep and love, certainly, but also for day-dreams, inspired by electronic images. Non-intrusive technology was blended into the decor of this total work of art, with media incorporated into luxuriant surroundings.

In another scenario, new products were seen as purveyors of a state-of-the-art experience at the centre of the domestic environment. Ron Arad's creation was an electronic gallery, a media room offering an intense encounter with pictures, where paintings by the Israeli artist Gavriel Klasmer appeared both in their original form and in electronically generated versions. The room was, in effect, a private museum, a space dedicated to the imagination and to meditation, with video screens forming a dynamic mural, a vehicle for ever-changing exhibitions. Whereas we might once have flicked through the pages of a book or magazine, in this room we could now change the painting by the press of a button.

Yet another example from the same exhibition was the Home Office, developed by Philips Design. This was a media room, an open house, where work is fully integrated with leisure. A flying carpet with laptop computer took us on a tour of the various spheres of activity engaged at different times of the day – morning, noon, afternoon and evening – using all sorts of electronic media. New interactive communication and information systems, ranging from telefax to picturephone and personal communicator, were available to provide the freedom we need to be able to work and play without structural restrictions.

The Milanese designer, Matteo Thun, integrated screens and hifi sound into a fantasy bathroom world. A world peopled by technological creatures resembling comic-book figures, with 'eyes' (LCD screens) which light up, and 'voices' which speak through loudspeakers. On the wall of monitor screens, waves came lapping right up to us, moving images as elements of interior design. Every corner of this fantastic room was penetrated by 'infotainment'. In Thun's words,

the bath had become a "new island in a new room", with "high-tech covering the surface of the room and the objects like a sensitive skin".

Thun's bathroom creatures are beginning to come alive, and the other media rooms, too, are closer to becoming reality than some of us may think. The common factor in all these imaginative worlds is that they involve products that add dignity to the home, that make it a place of pleasure, entertainment and happiness. They serve to create a warm environment, a place of communication and information, serenity, calm security and self-realisation.

It is possible, then, for us to improve the individual's 'culturosphere'. We can turn the home into a radiant landscape. But for this to happen, we must cease to think of the product as an end itself. Rather, it must be a creator and carrier of knowledge, services and emotions. Multimedia interactive products, for example, are important not only for the engineering that has gone into their creation, but also for the emotions which their images, information, music, and so on, can inspire in the person who uses them. In this shift towards viewing products as carriers of new qualities, design is moving from 'hard' to 'soft', from quantity to quality. Products no longer convey image, but identity. In the new world of the unlimited market, we must endeavour to create an environment that will encourage the process of cultural fulfilment.

**The other half**  At the same time, we must not ignore the non-Triad countries or, indeed, the poorer sections of the Triad countries themselves. Much of the world's population lives in poverty, and daily life is devoted to satisfying very basic requirements. In the developing countries, the consumer has yet to reach saturation point. Such countries still need to grow in quantitative and qualitative terms. Yet, as these countries endeavour to 'catch up', we must try to help them to learn from the mistakes that have been made in the Triad countries. Here, somewhere along the way, the relationship between the consumer and the product has gone wrong. People have become disaffected with products that are overdesigned and overcomplicated; and short and wasteful life-cycles have merely enlarged this sense of distance. The old affection for products has been lost: instead, the dominant feeling is alienation. We must ensure that the same thing doesn't happen in the non-Triad countries. In the industrial Triad, we must use available technology to restore the relationship between the consumer and the product. We need to use available technology to restore the relationship between the consumer and the product; and at the same time we need to make sure that the non-Triad countries don't get themselves in a situation where such a 'rescue operation' is necessary. We must seek to create a design strategy which will improve the quality of life in *both* types of society.

Clockwise from top left: Fantasy Bathroom, Home
Office, Domestic Media Room

Relevant products speak to real needs: an interactive
family tree, for instance, meets people's need to feel
they belong (from *Vision of the Future*)

**The race against time**   Modern times have been dominated by speed and acceleration. The implications of this have been enormous and diverse. Through the acceleration of the processes of production and consumtpion, the amazing acceleration of time in recent history has led to the explosive problem of the environment, to our increasingly superficial and banal relationship with objects, and indeed to the whole cultural crisis our society is experiencing.

But from crises we can look to opportunities; we must therefore reconsider the value of time. We must go beyond the myth of velocity. The entire complex machine of production and consumption must be slowed down. We have to take time to enter into a relationship with people and things: time to do things carefully, considering the implications of our choices; time to think, to contemplate, to savour; time to give meaning to our lives and what we are doing.

As Ezio Manzini, Director of the Domus Academy, says, "We live in a world of products that require very little effort and only minimal attention – a world of disposable products, made up of objects and images that slip right past us without leaving the slightest lasting impression on our memories. Objects that clearly minimise the effort and attention demanded but which, at the same time, produce minimal levels of quality and an enormous mass of waste and detritus." In opposition to this impoverished and wasteful type of relationship with objects, we need to shift our objectives. We should begin to work to achieve not, as before, a minimum of effort, but a maximum of quality; a quality of relationship that requires attention and care.

**Relevant objects**   To achieve this, a radical change in approach is necessary. If we're to provide relevant, honest and reliable products for both the developed and the developing world, then we must promote relevant, honest and reliable design. This also raises the question of ethics, a concept that has been sadly ignored in recent years. No matter who our consumers may be, we must try to give them relevant value for money. We need to abandon our obsession with adding extra functions or fancy gadgets to products. Instead we must enhance the quality of the consumer experience by making products easier to use. At the same time we must give back products their old dignity. Why, for example, does wood age gracefully, while plastic doesn't? It probably has something to do with the all-too-brief life-cycle of today's products. We need to replace the 'use-and-throw-away' mentality. We must make products more 'user-friendly'; not just by making them easier to use, but by restoring the traditional sense of friendship between consumer and product.

Caring for things means, as Ezio Manzini says, "considering them as creatures produced by our spiritual sensibilities and by our practical abilities. Creatures that, once they have been produced, exist and have lives of their own. Creatures, however, that need us as much as we need them." This entails a radical transformation of our points of refer-

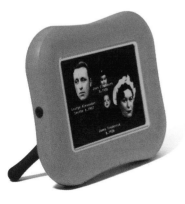

ence, our values and the qualitative criteria with which, until now, modern culture has evaluated the relationship between humans and their environment, between people and objects. It entails, too, a profound change in the culture of design. This change has been going on for some time and it has already generated the complex phenomenon which has taken on the name of 'New Design'. For New Design, the relationship with objects is placed at the heart of a wide-ranging meditation, offering a springboard for a great many lines of research.

As a part of this strategy, Philips gave technical and cultural support to the *New Tools* project, presented by Marco Susani and Mario Trimarchi at the Triennale di Milano, 1992. The New Tools were concepts for small electric appliances of the future, for the kitchen and the table. They were based on the idea that the 'machines' used in the kitchen should be more congruent with the culture of cuisine and gastronomy, and with our behaviour in the kitchen. Designed to be like the kitchen equipment of old, without the usual disturbance caused by high-tech machinery, they had solid-state components and mechanics, making them quiet and gentle. Their interface was natural, and they were not tied down to an external source of energy. In all respects, they were designed to be infinitely far removed from the 'white plastic look' we are so familiar with.

**Interdependency**   If we're to address the high complexity the world presents us with, the strategies we use to combat it must be mutually supporting. In fact, there's a direct correlation between high complexity and interdependency. No company these days can develop a product or a process without the support of another. A car manufacturer, for example, is dependent on the chemicals firm which provides his materials, and vice versa. Similarly, within a single company, there is interdependency between different competencies, such as engineering and marketing. Developments take place simultaneously; the linear model has vanished. New products must be compatible with those that came before and with those that will follow.

In the face of today's 'hyperchoice', information is vital if we're to cope at all. Everything is possible and available, and effective cooperation and communication is therefore essential. This applies just as much to design as to other fields. In today's complex world, it's no longer possible to tackle a design problem from the point of view of a single skill: a multidisciplinary team approach is required.

**High Design**   The answer to High Complexity may be sought in what I call 'High Design'. By High Design, I mean an integrated process incorporating all the skills on which design has historically based itself, plus all the new design-related skills we need to be able to respond to the complexity and challenges of the present and anticipate those of the future. The High Design process is one which continuously adopts more advanced cultural and technical criteria. It's based on the fusion and interaction of high-level skills.

Certainly, calls for the collaboration of designers, psychologists, ergonomists, sociologists, philosophers and anthropologists have been made in the past (Professor Misha Black's address to the ICSID Conference in 1972 comes to mind). But they usually had little practical result.

Design in a world of high complexity should no longer be a case of clever individuals or teams creating products in splendid isolation, but of multidisciplinary organisations or networks creating 'relevant qualities' and 'cultural spheres'. If we're to make the quantum leap from the limited materialistic and quantitative market to the unlimited, more spiritual and qualitative market, then we must provide the design worthy of it.

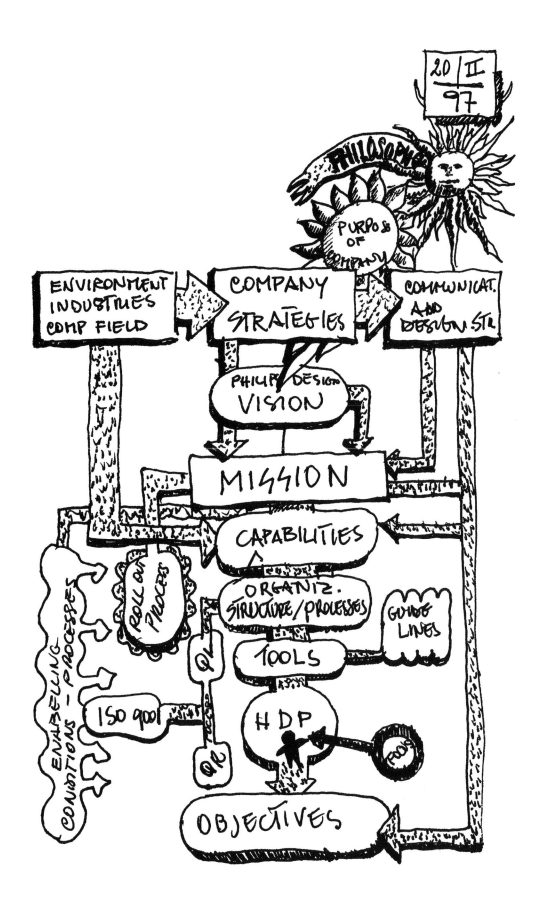

**Etica nova** In order to meet all the challenges facing us as a community, we have to adopt sustainable strategies; we have to adopt High Design processes. But more than anything else, we have to create and adopt an *Etica Nova*: a new ethic as the driving force behind all we do, an ethic which incorporates universal values such as Love, Brother/Sisterhood and Peace. We can also add Ecology to this list – not merely in the sense of 'environment', but in its deeper meaning of 'harmony between the elements', both in the artificial world we construct and in the natural world around us.

These noble sentiments may appear incompatible with the concerns of a modern, profit-making company; and here lies our greatest challenge. We're working for a large company, and, at the same time, are concerned about the quality of our lives and those of our children. But there need be no contradiction here. We're in a position to convert the company we work for to a more ethical design philosophy; and we can do this by making the quality of life a competitive issue. Products must compete in terms of their capacity for improving the quality of people's domestic and working environment and for promoting their cultural fulfilment. It is this, rather than quantitative, redundant gadgetry, which must become the key to the future.

The technology we need to do it is available. In fact, technology is not the challenge: it is the way we apply it. We must use technology as a force for 'Good' rather than 'Evil'. It is possible: we *can* create a 'landscape of happy objects', a harmonious, peaceful Paradise Regained – rather than the materialist 'hell' of modern Las Vegas.

Tradition and innovation: quiet, cordless kitchen
appliances reflect trusted models from the past
(concepts from *New Tools*)

PER RITARDATO INCONTRO VENEZIANO

21|81 94

ITALIA

SOCIO CULTURALE

OPPORTUNITÀ TECNOLOGIA

VISIONE — INNOVAZIONI

INTER

INNOVAZIONE UOMINI E OGGETTI

ARCO DELLE XXIV ORE

TEATRO · DELLA · NARRAZIONE

UNA IDEA PER UN NUOVO TEATRO DELLA VITA . UNA IDEA DI 24 ORE , SCAMPOLO DI VITA , RESPIRO DI UNA ESISTENZA ... IDEA DI UN ATTIMO , TEMPO VERO UNICO OPPOSTO DELLA ETERNITÀ... UNA IDEA DI CONTINUITÀ QUINDI DI CAMBIAMENTO UNICA VERA COSTANTE DELLA STORIA. UNICA CONTINUA SORPRESA UNICA CONTINUA PREOCCUPAZIONE DEGLI ADULTI , UNA VERA ATTIVITÀ DEI GIOVANI... RUOLI DEL DESTINO? TRASFORMAZIONI PREVISTE - PROGETTO DIVINO?

# The Smile in My
# Grandfather's Mirror

Grandfathers are wonderful people. I was very fortunate in mine, and I learned a lot from him. He was a tailor in Italy and worked at home. We had a great relationship, and I spent hours playing on the floor in his workshop. He didn't turn me out when customers came (I suppose I must have behaved myself!) and I saw how people came in, looking for a new suit or a new dress. In those days, you only bought new clothes for a very special occasion, such as a wedding or christening. People usually didn't really know exactly what they wanted. All they knew was that they wanted to look and feel good! So an important part of my grandfather's job was to find out what sort of clothes would make them feel like that. He'd show them books of drawings with different styles of suits, showing people in various scenes and poses. There was talk of double-breasted or single-breasted, the number of buttons, the shoulders, the sleeves. And then there was the cloth – the material, the pattern, the colour. It was a process of stimulating and refining, of imagining and dreaming and perfecting, getting closer and closer to that image that was still indistinct, but getting clearer all the time. It was a question of trying to meet the great expectations that every customer brought into that room.

Then, when the customer had left, grandfather began to apply his skills: drawing the pattern, cutting the cloth, making the first seams. His years of experience came into their own as he put into practice the traditional skills of his ancient tradition. Several weeks later, the customer came for the first fitting. It was still difficult to see what the suit would be like in the end, but it was a phase that had to be gone through if the final result was to have the desired effect. Finally, the moment arrived. The suit was ready. It was a very special moment, and a serious one, because the special occasion was only a few days away. The suit had to be right; there would be no time to make another. The customer took off his normal clothes and put the suit on in hushed silence – the silence of expectation. Finally, he turned to the mirror. By this time, sensing the solemnity of the moment, I had emerged from under the table, and grandfather and I stood side by side, watching for the customer's reaction. Then, as he looked into the mirror, and saw the dream become reality – the smile, that smile that spread across the customer's face! It was truly a moment of magic.

**How to become the customer's first choice**   Of course, when a customer smiled like that, my grandfather knew he would almost certainly return on another occasion. If we're to get our customers to smile like that, if we're to provide them with things that realise their dreams, create meaningful objects, communicate with them and inspire them with emotions – emotions of pride, confidence, and self-worth. And if we want them to come back and ask us for more, in other words, if we want to become the customer's first choice, we need to understand them. And we need to understand them well. At Philips we're trying to get to grips with the future by trying to understand today's consumer, so that we can make sure our products and services will be the customer's first choice. My grandfather had learned by years of experience how to do it for his customers. But today's global customers are complex beings, and we don't often get to talk to them on a one-to-one basis. That makes our task even more of a challenge.

**Meaningful products through looking ahead together** My grandfather knew it was no good making a suit that just looked flashy. It had to fit perfectly; it had convey the customer's ideal of quality. And the same goes for us. It's no use providing products that are only superficially attractive. They also have to reflect the user's deeper values. To survive, any company has to make sure its products and services are the customer's first choice. And they will only be that, if they respond to what the customer really wants – something that has meaning.

How, then, can we provide those meaningful products? How can we look ahead and see what people will value and also find attractive in five to ten years' time? The only way is to involve people – working together with others. Only a vision of the future that's based on many different views has any chance of being both attractive to a large number of people and technologically feasible. History has shown all too often that visions of the future created by single individuals tend to turn out laughably wrong. The future is apparently too complex to be foreseen by the limited mind of one person.

The problem, too, is that actions that seem to promise improvements in the future can have unexpectedly mixed consequences. Take, for example, the vision of the Modernist movement earlier this century. Through mass production and relentless industrial progress, it promised greater welfare for more people, but, although it certainly brought many benefits, it also produced pollution on a scale we are only now beginning to understand. Similarly, in agriculture and food production, the efficient ways of working that have done much to relieve hunger, limit disease and save time, have also been responsible for over-worked soil, unhealthy fast food, and mad cow disease.

**Focus on customer benefits** Now if we, as designers and manufacturers, are to take advantage of this situation to do our bit to help in the attainment of a sustainable society, we have to genuinely pay attention to consumers. We have to make sure that what we do fits in with what they want. And what consumers want is not products, but benefits. We therefore need to shift our focus from products to customer benefits. We need to think not of a product creation process but of a customer benefits creation process, with the customer at the start and the end of the process.

Before we start developing a product or a service, we have to answer questions like: What actually benefits people? What do they consider meaningful? What does 'better' mean to them? What delights them? What in a product attracts people so much that they can't help but choose it above all others? If we can find answers to these questions, and translate them into products and services, there's a good chance we shall be on the right track.

**Understanding the customer** This means we have to understand as much as we can about our potential customer. Customers have many dimensions and play many roles. We need to get to know them in all these roles: as user, owner and buyer. We need to know what sort of things people find pleasing to use, own and buy; what sort of things arouse feelings of pleasure and pride and the sense of meaningful activity. We also need to find out what it is that makes products attractive enough that people want to pay money for them, and for them rather than all the others on the market.

In other words, we need to understand much more about human psychology and motivation. Technological expertise and insight are not enough. They need to be balanced with an equal level of 'people expertise'. The potential customer is a multidimensional human being. By studying people with respect to all these characteristics and dimensions, and learning from them, we can set up a process which will allow us to create and present products and services that the customer will find attractive and desirable, as well as useful and easy to use.

**What does the customer want?** So is being 'customer-driven', 'customer-focused', 'customer-oriented', and so on, really all a company needs to become the customer's first choice? It would be so easy if

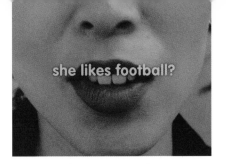

we could just ask people what they find attractive. But, unfortunately, that won't help. What really delights people is when you give them something nice that they hadn't expected. And this may be something that they didn't expect simply because they had never even thought of it; or if they had thought of it, they never believed it would be possible.

People are notoriously unable to forecast what is possible. Akio Morita, former chairman of Sony, put it like this: "Our plan is to lead the public with new products rather than ask them what is possible. The public does not know what is possible, but we do. So instead of doing a lot of market research, we refine our thinking on a product and its use and try to create a market for it by educating and communicating with the public."

People also have little idea of what they want until they actually see it. So we need to find out for them. Prahalad and Hamel, writing in the *Harvard Business Review,* expressed it like this: "Some companies ask customers what they want. Market leaders know what customers want before customers know it themselves."

**A vision of the future**    So if we can't ask customers about their ideal future, we need to find ways of looking ahead and seeing it for them. We need to develop the best possible hypothesis about the future and work with ideas on that basis. In other words, we need foresight. We need to formulate a vision of the future, i.e., a hypothesis about what is likely to happen. This is not only important so that we can provide people with the benefits they really want; in practical terms, it will also help us to answer three critical questions: First, what new types of benefits should we seek to provide in the next 5, 10 or 15 years? Second, what new competencies will we need to develop or acquire to be able to offer those benefits to customers? And third, how do we need to reshape the customer interface over the next few years (communication, distribution, sales strategy, etc.)? A vision of the future, then, is essentially a hypothesis about benefits, competencies, and the customer interface.

A good example of where having such a hypothesis has worked well is that of Motorola mobile telephones. Motorola foresaw that people would like real person-to-person telecommunication rather than place-to-place communication. So they learned about digital compression, miniaturisation, batteries and flat-screen technology, and they developed their customer interface and brand familiarity to be able to sell this idea.

Another example is that of Apple computers. Apple's success story comes from their prediction that everyone would like to have their own computer. The problem was that people found using computers difficult. So Apple developed its graphical user interface, giving them their excellent reputation for user-friendliness.

Thinking about the future in terms of benefits, competencies and the customer interface allowed these companies not just to create successful products but to create whole new worlds of possibilities – worlds in which they took the lead.

**'Imagineering' the future**    These examples show that, to get ahead, what we need is a well-articulated vision about tomorrow's opportunities and challenges, based on a thorough understanding of trends in a wide variety of fields. But it's not enough just to collect information. The crucial step is to integrate that information into a process of imagination and prediction, a process that will allow us to construct a powerful visual and verbal representation of what the future could be like: making the future come alive, engineering the future through imagining – what is sometimes called 'imagineering'. There are three things we need to do to facilitate this process.

**Not business units, but competencies**    First, we have look beyond the market we already serve. How can we do this? One way is to change the way we look at companies. We tend to see them as a collection of business units. Instead, we should see them as a portfolio of core competencies which can be grouped and

Left Limited edition Swatch watches for the global traveller, designed by Stefano Marzano

Opposite, below Making the future real: the Philips-Alessi line, *Television at the Crossroads*, *Vision of the Future* and *New Objects, New Media, Old Walls*

regrouped in many ways, lots of them totally new. Canon, for instance, combined the functions of a scanner, an optical recording device, a keyboard, a monitor and a printer to produce a desk-top document storage system – a new product with a new market.

**Not products, but needs and functions**   The second way we can facilitate the process of imagineering is to stop thinking in terms of traditional products and services and focus instead on needs and underlying functions. Oki's development of the electronic whiteboard is a good example of this. They combined their competencies in scanning and printing to come up with this new product. They looked beyond the form of the traditional whiteboard, and focused instead on its functions. What do people want from a whiteboard? Can their needs be met in any other way? Can other, hidden needs, be met at the same time? Oki saw that the answer to these questions was 'yes'. And the result was a product that can do everything the traditional board can do, plus being able to preserve and print out what has been written down.

**Ask naive questions, hope for the impossible**   The third way to facilitate imagineering is to ask naive questions and hope for the impossible. If you're going to predict the future, you must be willing to ask, "Why can't it be different?" It was a simple question like this that led to the invention of the Polaroid camera. When Edward Land took a photograph of his daughter, Amy, she asked why she couldn't see it at once: that set him thinking and the result was instant photography. In the same way, when Nicholas Haytec of Swatch noticed that people don't only wear a watch to tell the time but also to express style or status, he went against Swiss tradition by asking, "Why don't we make a cheap watch which will let people make a statement?" And that opened up a vast field, because the range of possible statements people might want to make is infinite.

**New opportunities where trends intersect**   Predicting the future therefore requires more than just gathering information. We have to look beyond what we're used to and shake off our traditional inhibitions. When we've done that, we're ready to start the real imagineering process: analysing trends of all sorts and imagining ways in which they might interact in the future so as to create new business opportunities. There are some interesting examples of successful products which are the result of this step. In each case, someone foresaw how various changes which were on-going or just starting came together to create a great business opportunity. Take the personal communicator. This product lies at the intersection of a variety of changes in lifestyle (incessant travel, more information to be processed), technology (miniaturisation, digitalisation, digital compression), and regulation (the freeing-up of additional bandwidth). The changes in lifestyle created the need, the technological changes made it possible and the changes in regulation meant that it could implemented.

Another example is CNN, the 24-hour global television station. This service lies at the intersection of a different set of changes: people working longer and unpredictable hours, the advent of handicams and suitcase satellite link-up, and the licensing of satellite and cable TV companies. Again, the changes in lifestyle created the need, the technological developments meant that it was possible, and the changes in regulations meant that it could be implemented.

**Leading the customer** The companies who came up with these products and services were giving shape to a need that had not be spoken. People didn't know they wanted these things, but when they saw them, they liked them. The lesson we learn here is that companies and designers do not serve people well by waiting to be told what to do. They often have to take the lead, based on their informed intuition of what people want.

**Making the future real** Once we think we have clear ideas about future benefits and future customers, how can we test them? Basically, we have to try to stimulate people into expressing or articulating their subconscious 'dream', their latent needs and desires. Our hypothesis about the future, based on research into human psychology and into social and technological trends, is an abstract idea. What we need to do is to make it tangible, visible, *real*. If we do that, and show real objects to people, we can get them to react, to tell us how they feel about it, what they like and what they don't like about it, how they would improve it, and so on.

This approach of showing people abstract ideas in tangible form is one we at Philips are pursuing in earnest. Its main objectives are to gather knowledge about people and their sensitivities, and to generate new business opportunities. By showing people abstract ideas in realistic form, you get feedback. And from that feedback, you can derive a lot of useful information and knowledge about their subconscious wishes and wants, all of which would otherwise have remained hidden. By showing people something concrete, you sow in people's minds the idea that all sorts of things might be possible that they hadn't thought of before. Seeing unthought-of things triggers a mental reaction. Something that was, until then, perhaps only a germ of an idea deep in their subconscious is moved up several levels towards the surface. From being no more than a vague dream it moves towards becoming a desire. It is then only a short step to it becoming a psychological need. By triggering this process in people's minds, you are sowing the seeds of new business: one day soon, people may want a product like this.

**Strategy in practice** Over the past few years, we have put this strategy into practice in a number of projects. The Philips-Alessi line of kitchen appliances was the result of a project to explore developments in the kitchen environment. Our project *Television at the Crossroads* (1994) explored the merger of television and the computer.

Personal Helpers: electronic tools integrating
communication, entertainment and information-
gathering (concepts from *Vision of the Future*)

To explore various possibilities, we designed concepts which presented the television set in various forms, ranging from tele-lamp, tele-table, tele-hearth and so on, right through to a tele-pet. We showed television sets carrying various services and functioning in various ways, such as tele-shopping, tele-playing, tele-partying, tele-monitoring, tele-mirroring, tele-relaxing, and so on. Basically, the exhibition suggested the potential advantages of 'being connected' with the world from your own home. *New Objects, New Media, Old Walls* (1995) involved objects that were a step ahead of both the TV and the PC and explored the problems of integrating advanced technology into the home in an acceptable way. Our most extensive venture into future-projection to date has been *Vision of the Future* (1996), an attempt to stimulate a wide-ranging discussion about the sorts of products and services we should be offering the public in the coming decade.

Our initial creative efforts for *Vision of the Future* were preceded by a careful analysis of emerging socio-cultural trends around the world. The world today is characterised by new understandings of what time and space involve. For instance, many businesses are conducted not only around the world but also 'around the clock' – there is no natural break in their business day as there used to be. In the same way, with the advent of cyberspace, we now understand space differently: geographical distance has become less significant than it used to be in our lives. Now, these new developments interact with others. Take work, for instance. In the future, we shall not need to travel to work so often but will be able to work from home. Thanks to telecommunications and computer technology, we have overcome the problem of the current geographical separation of the locations of our two primary activities, work and relaxation. Add e-mail to these technologies and you have the possibility of working satisfactorily with someone else but at different times, not necessarily during traditional office hours. These are trends which are already becoming visible. More interesting, because less obvious, are potential or latent trends in human behaviour, psychology and culture, phenomena which we have called 'sensitivities'. They're the outcome of a social 'chemical reaction'. They may (or may not) 'take off' and develop into trends. An example is the sensitivity (in our sense) of people getting together in virtual communities via the Internet. In the past, we could only form groups with those who were in the same geographical area as ourselves, but, as people get used to the idea of contacts in cyberspace, this is beginning to change. In developing *Vision of the Future*, we considered many of these trends and sensitivities and tried to match them up with appropriate trends in technology. One of these technological developments was the ever-increasing miniaturisation of products and the increasing computing power afforded by the silicon chip, which is making voice recognition and voice synthesis a practicable way of interacting with products. In the world of software, we're seeing the development of software agents, programs that are able to think for themselves – or rather, think for us. They will be like a small community of personal 'helpers'.

Another technological trend we looked at was that of 'smart' or 'interactive' materials. These are materials that can modify their behaviour under specific circumstances, changing their shape, stiffness, position and so on, depending on temperature or electro-magnetic fields. There are also plastics that can change colour or opacity and then revert back to their original state. And finally, nano-technology (micro-electronic techniques developed in chip technology) is making it possible to create tiny sensors. Many people already have one of these in their car: it decides when to inflate your air-bag. But they can be further developed to detect even more subtle things, like smells.

This sort of information about emerging socio-cultural and technological developments formed the input to multidisciplinary workshops. There we explored how they might interact to give rise to new products and services. These workshops resulted in some three hundred new proposals. Then, using certain criteria, we filtered these 300 ideas down to only 60 for further development, and a panel of international experts (futurologists, sociologists and trend analysts) were invited to comment on them. The next step was to refine the concepts and begin to transform them into realistic models and simulations. This not only involved the physical appearance of the product but also how it could be used in real life. We therefore developed hard models and made short video clips to show the product being used in realistic future situations. The results were communicated to a wider public through an exhibition in Evoluon, the Philips Competence Centre at Eindhoven. We also produced a book, a multimedia program and a Web site to spread the ideas and generate feedback.

**Strategy for growth**   In this way, we're trying to help our company achieve its goals by pursuing a very specific strategy. It's one that allows us to find new customers and non-traditional products by finding out about people's unarticulated needs, needs they have or will have in the near future but cannot describe. Our strategy puts the customer at the centre of a customer benefits process. It's part of our way of 'making things better', and, I believe, the way to bring that same smile to our customer's face that I saw in my grandfather's mirror all those years ago.

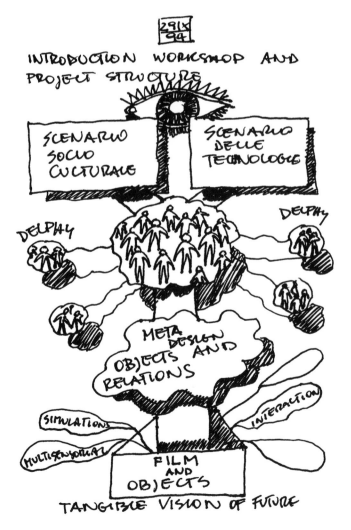

The design process for *Vision of the Future*

# Into the Era of Soul

**The metaphysics of television**   What is the nature of television? What are its first principles? I don't mean here its technology but rather its ultimate *esse*, and its deeper significance in our lives. Although no easy task, we certainly need to explore such questions if we're to design televisions appropriate for the twenty-first century. The way we interact with television is becoming increasingly complex and the need to improve our understanding of the medium is therefore all the more urgent.

All artefacts, I suggest, may be viewed as having two essential elements. Let's call them, by analogy with traditional metaphysics, the Body and the Soul. The Body comprises those characteristics of the artefact which give rise to nothing beyond themselves. They're limited, bound in time and space. They gain access to the human faculty of reason, but cannot penetrate the human imagination. The Soul, by contrast, comprises those aspects which are intangible and unlimited. By appealing directly to the human imagination, they trigger an infinity of ideas and emotions. Transcending the physical, they speak to the soul of the beholder.

In these terms then, a traditional object, such as a statue or a chair, can be said to have a Body (its physical form) and a Soul (the meaning, the associations and emotions, aesthetic or affective, aroused in the viewer or user). Television, however, in common with other products of high technology, has a more complex metaphysical structure, corresponding to its complex physical structure. At first sight, we might say that the Body of television is its hardware, i.e., the electronic circuitry, the screen, the controls, and the casing required for the device to operate effectively. Its Soul would then be its 'software', the programmes, the sounds and images which entertain, inform and inspire the viewers. Each, thus defined, is clearly a typical example of its category. But such a structuring is too simplistic. The situation is complicated by the fact that a television has two modes of existence. It can be switched on or switched off.

When the television is on, our attention is directed exclusively towards the software, which, as we've said, can be seen as its Soul. Just as in a theatre, when the lights go down, all eyes are concentrated on the stage. The software is performing. We can be inspired by the stories it tells, we can be thrilled by the sensory input it offers us. But we can also identify a Body of this software: the form of the graphics, the style of the photography, the quality of the acting, and so on.

When the television is switched off, it is the hardware, the Body, that comes into its own. In its way, it too performs and displays its own Soul. The appearance of the set itself has meaning, communicating certain cultural and aesthetic values which have the power to inspire and delight. Again, the comparison with a theatre is not entirely inappropriate, since the decor of the auditorium and foyer enhance our experience of the play by heightening the sense of occasion and expectation as we wait for the curtain to rise.

With the tremendous expansion of possibilities provided by telecommunications, satellite broadcasting and multimedia, which expand the realm of the software, it's all too easy to push hardware, the carrier of the software, out of

Below: Homes are becoming polycentric, with multimedia integrated throughout (stills from *Vision of the Future* videos).

the limelight. But, of course, the hardware is necessary for television to exist at all. The software cannot achieve its end – the inspiration of the viewer – without the cooperation of the hardware. Similarly, the Body, without its immanent Soul, the software, loses its primary reason for existence. There is, in other words, a relation of interdependency between the two.

**The polycentric home**  A second issue we need to address is the way television relates to and functions in the domestic space, the home. For several millennia, the development of domestic architecture has followed essentially the same pattern. It has been characterised by the progressive implementation of the concept of having separate areas devoted to separate tasks or functionalities. From an original single, open-plan living space, in which all activities took place, including the care and housing of domestic animals, the home has become increasingly subdivided by walls into monofunctional areas. The kitchen for cooking, the bedroom for sleeping, the dining room for dining, the bathroom for ablutions, and so on. In the same way, the fire, formerly a source of heat for cooking, of warmth for comfort, and consequently also the location of social activity, has now been dispersed around the house: fire for cooking to the kitchen, fire as social focus to the living room and fire as source of warmth throughout the house by means of central heating.

Although our domestic architecture still adheres to the traditional concept of monofunctional areas, there is now

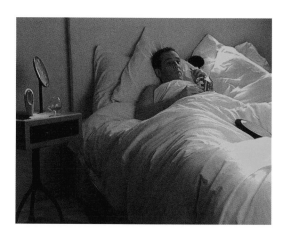

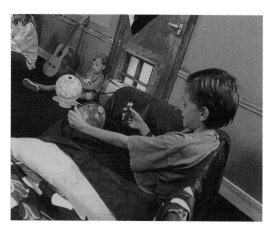

overlaying it a new, virtual matrix. This new matrix is a remapping of living 'space' based on leisure and social activities, rather than on functions. The home has become polycentric; rooms have become multifunctional. One member of the family may be playing with the computer in the bedroom, another may be watching television in the study, a third may be reading in the living room and a fourth telephoning in the kitchen. They may eat formally in the dining room or informally in the kitchen, in the sitting-room on a tray, or have breakfast in bed. We can prepare food in the kitchen, cook fondue in the sitting room, or barbecue in the garden.

This development is a reflection of the increasing fragmentation and individualisation of industrial and post-industrial societies. People today are more individualistic, no longer following clearly pre-defined patterns of behaviour. No more do children automatically follow in their parents' footsteps. Families are becoming communities of individuals, just as consumers in general are moving away from group style towards individual style. They're making their own choices, creating their own 'culturosphere', rather than simply adopting the values of others.

Both this process of individualisation and the evolution of the home into a polycentric environment have affected the role of television in the domestic setting. Initially, as a newcomer, television was assigned a peripheral position in the home, the corner of the sitting room. Despite this lowly status, it quickly became the new point of focus for domestic life, the source of entertainment and information. Psychologically at least, it moved to centre-stage. In this respect, it usurped the role of the hearth, around which tales were once told and exploits recounted. The televi-

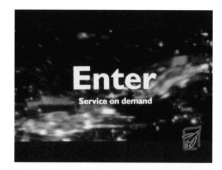
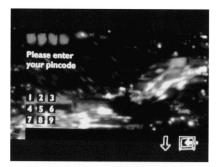
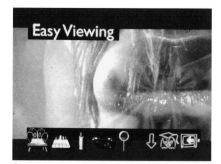

Interactive Services-on-Demand systems allow viewers to access entertainment or information services at any time

sion took over the parts of story-teller and minstrel alike, and provided the flickering light and psychological warmth that made us feel truly at home.

But just like the fire, physically, the television is now being dispersed throughout the house – to the bedroom, the kitchen, the study and maybe even the bathroom. We can watch the channel we like, at the time we like, in the environment we like. The participation of television in the restructuring of the home from a monocentric to a polycentric environment means that television is on the eve of a radical transformation. It will develop beyond its present virtual monomorphism to become truly polymorphic. It will cease to be monofunctional and will take on new roles. In short, it will start to do and be exactly what we want.

**Domesticated technology** Technology was once an alien in our familial scene. The television was initially housed in a wooden box and placed uneasily in a corner. Now, however, such technology is rapidly becoming domesticated. We now feel that many of the technological products that share our homes are as dear – or at least as significant – to us as pets or friends. There are many more such 'friends' just waiting to be invited in to join us, and television is likely to be the focus for many of them.

The new personalised settings in the polycentric home will give rise to new forms of the set itself. Until recently, sets varied little in their general characteristics. With few exceptions, they were not designed to fit precisely into the setting in which they would appear. But as the number of televisions per household increases, there will be a greater need for different-looking models. The decorative quality of the hardware will come to play an increasingly important role, expressing its own personalised visual character.

In the bedroom, for example, we'll curl up intimately with our mini-screen TV, children will watch it secretly under the bedclothes or talk to it like a teddy-bear. In another room, we'll create our own wide-screen cinema – basking in an all-enveloping sensory experience, enjoying the sense of *being there*. And in another area of the home, the television may also, as I hinted above, fulfil the role of *objet d'art*. It may inspire and please us aesthetically simply by virtue of its outward appearance – what I referred to above as the Body of the hardware. In short, we shall all want our own set to express our own personality and satisfy our own individual needs. But the needs that can be met will go far beyond those that television technology is currently able to address. Technologies are converging fast. Already television, telecommunications and computer technologies are being integrated. Others will undoubtedly follow. This development will lead to a whole new range of functions and a whole new range of needs that they can satisfy.

**Interactivity**    One of the most significant consequences of the fusion of various digital technologies will be interactive television. The effect of television over the past few decades on social intercourse has been dramatic. But it's an effect that has been something of a two-edged sword.

On the local level, television initially resulted in a degradation of social interaction. When people started staying at home to watch television instead of going out to the theatre, the cinema or the sports stadium, the group experience became an individual experience. Once we used to be in the physical presence of performers and players, though we related to them only from our status as members of a crowd. Now, on television, often alone, we view an image in only two dimensions, but can nonetheless enter into a quasi-intimate relationship with the people on the screen. At the same time, the family changed from being a genuine group, engaging in mutually stimulating activities, to a group of individuals, physically present together, but, as they watch television, each individually involved in a virtually exclusive relationship with the screen. Locally, then, television has led to a deterioration in the quality of social intercourse. The relation between the members of the family or social group has changed. Stimuli are now sought outside the family circle, and not on a group basis but alone.

On the global level, however, television has expanded our horizons enormously. Combining as it does the two fundamental senses of sound and vision, it allows us to perceive and experience, in more realistic ways than ever before, parts of the world we're perhaps incapable of reaching physically. Our social awareness on the macro level has been raised to

Small movable cameras provide an extra eye on the world (from *Television at the Crossroads*)

heights previously unknown. The vast array of images to which we now have access also gives us similarly increased access to the immaterial world, inspiring fantasies and engendering dreams.

Thus television seems to have diminished two-way communication on the local level, while enhancing one-way communication on the global level. But we're now about to enter the era of interactive television. As we move from passive watching to interactive televiewing, we shall at least be moving towards integral social communication at global level, a development which may go some way towards restoring the balance between social and individual activity, between communication at the local and at the global level.

Certainly, one of the social benefits that television has provided has been that many people have been able to experience human contact which they'd otherwise have been denied. The elderly and housebound have been able to remain informed about the outside world; the lonely have felt they have had company. But so far this contact has only been passive; people have been placed in the role of voyeurs, powerless to exert any influence on what they observe. Now, with the coming of interactive television, they'll be able to participate in more genuine social interaction through this medium. In addition, with the increasing individualisation of telecommunications and television transmission, people will be able to communicate interactively with loved ones far away. Young people will be able to be present at happenings of their choice all over the world. Those who for one reason or another are unable to travel will be able to enter and experience other cultures.

**Expanding experience**   However, although interactive television is arguably the most significant consequence of converging technologies to emerge so far, it is not the only one. Let's just speculate for a moment on a few more intriguing possibilities.

We all cherish moments and experiences shared with loved ones: school friends, first loves, happy holidays, children's smiles, and parents or grandparents no longer with us. What is now represented by photograph albums, home videos, or that bundle of love-letters tied up with ribbon could soon be integrated into a digital 'home treasure-house', an ever-accessible visual and aural record of our history, an aide-mémoire to help us get more from our memories and that special collective experience which is family and friendship.

Television is in a sense an extension of our eyes, but we have no control over where the camera is directed. But suppose we add an extra 'eye' to our television, a small movable camera? Those with failing eyesight could use it to magnify writing or printing, with the enlarged text being displayed on the television screen. Elderly people or the house-bound

The TV mirror: watching the morning weathercast while brushing your teeth (from *Vision of the Future*)

used to sit by the window, watching the world go by. But that's not always easy in a modern high-rise building – unless you have an electrical eye on a flexible neck. They could use such an 'eye' to keep in touch with life just outside their home through the television screen. A camera linked to the television would do away with the need for mirrors, with their built-in disadvantage of lateral image reversal. And we could even wash, shave or make up, looking at ourselves in the bathroom 'TV mirror', while in a corner of the same screen we watch the latest news and weather on breakfast television.

In the traditional home, the fire not only provided warmth, but it also served as an object of meditation, stilling the mind and freeing the spirit. Gazing into a blazing fire, seeing forms come and go, thinking over problems, remembering loved ones is an important psychological activity, but one not always possible in today's centrally heated homes. However, television, suitably programmed, could easily take over this hypnotic function, helping to develop, as it were, 'the third eye' within ourselves.

**Beyond the horizon** As a result of the integration of various digital technologies, our reach, previously limited to our immediate environment, is being extended to the ends of the earth. We shall have access to a source of experiences far richer than we have ever known. Our power to know, to communicate and to influence will be greater than any generation that has gone before. It will be up to us to take intelligent advantage of the incredible opportunities such developments present for enriching and enhancing our lives.

# Designing Reconciliation

**Eternal irony** How ironic it is that as humans we long for a stability and certainty we can never achieve. Intrigued by the new, with its prospects of wider horizons and unknown worlds, we embrace it eagerly, knowing that, as we do so, we leave our former certainties behind. The price we pay is at best bitter-sweet nostalgia and at worst debilitating disorientation. We pay the price willingly to gain a new benefit; but the pain of uncertainty remains. Now, at the end of a century in which the steady march of technology seems to have broken into an undisciplined dash, and a new bout of uncertainty begins to assail the industrialised world, we need to take up the challenge: How can we ease the pain of leaving the past while opening new paths to the future? How can we reconcile old values with new opportunities, the reassuringly familiar with the disconcertingly new?

**The threat of technology** Through the ages, technology has played a major part in this recurring scenario. We are now witnessing the latest reaction. Nineteenth-century faith in the power of science and twentieth-century confidence in mechanisation and automation have begun – on one level – to be called into question. An environment poisoned by industry and the car, people poisoned by radiation disasters with nuclear power, mass unemployment as a result of automation, genetic engineering and the prolongation of life beyond normal bounds – such factors, seen as the result of uncontrolled 'scientism' and 'technologism', are causing many people to feel uneasy, even frightened.

At the same time, other achievements in the same fields are accepted unquestioningly as benefits, and there is still a belief that, as far as technology is concerned, anything is possible. But even there, we find a reaction setting in: not so much to the benefit itself, as to the effort required to obtain it. Many of today's products which bring pleasure and convenience into our lives are complex and difficult to operate. Frustration and alienation are often the result. Partially as a result of this, high-tech products, unlike the 'trusty tools' of old, fail to engender affection in the owner. As other uncertainties have increased, they have ceased to be seen as proud evidence of modernism and become an almost intrusive, alien presence in the domestic scene: bulky and black, or shiny and white, with trailing wires and glowing dials…

**A fragmented society** Increased mobility has meant that our society is more fragmented than in the past. It is now relatively rare for all generations of a family to live in the same town, let alone the same house; and the digitalised complexity of social organisation means that we are identified more often by a number than by a name. Market pressures have also meant that the friendly corner shop has been replaced by the impersonal supermarket. The local community is no longer a tight-knit group, the natural focus of its members' activities. Outsiders come and go before others in the community can get to know them. All these factors serve to erode our sense of stability, our sense of roots, and, ultimately, even our sense of identity.

**Transversal individuals**  This physical and social fragmentation of society encourages – and is encouraged by – a trend towards individualisation. Unlike their ancestors, people don't belong to a single tribe which is not of their own choosing. Instead, they belong to any number of groups, at least some of which they assemble themselves – family, friends, colleagues, people sharing the same hobbies, interests and values. The members of those groups may be geographically spread over a wide area. Some groups members they will know; others they'll never see, but simply identify with, by wearing certain fashion attributes. And these allegiances may change at a moment's notice, depending on mood, context or circumstances. People have become transversal and the homogeneity of the group is dissolving into the diversity of the individual.

**New domestic qualities**  These changes in society and the individual are reflected in a feeling of uncertainty regarding place and identity. We're beginning to witness, in reaction, a longing for a safe and intimate home base, one in which the traditional rituals can take place, in which socialisation and human warmth are again paramount.

Tradition and ritual provide us with a deeply reassuring sense of belonging, the confidence born of knowing one's roots. Our psychobiology is geared to an earlier mode of existence, one in which our physical and mental surroundings changed at a much more leisurely pace than we experience today. In comparison with our forebears, we are all refugees within our own lives. Driven by accelerating technological and geopolitical change from one temporary haven to another, we're forced to leave behind much of what we have known and loved as we move on, hoping for new opportunities, but apprehensive about whether we shall be able to cope with them. Home must therefore be a place where we do not feel threatened by high-tech, but where we can profit to the full from the benefits it provides. We need to feel comfortable with it, and experience it as a natural part of our heritage-filled home.

Over the past century or so, the distribution of activities in the home has changed. We have moved from the monocentric to the polycentric home, where particular activities are not confined to certain rooms but can be pursued almost anywhere in the house. This development reflects the increasing fragmentation and individualisation of society itself. Even families are becoming communities of individuals. Just as, during the past century, in the industrialised world, we've witnessed the dispersal of the extended family, so we now seem to be seeing the loosening of the bands which have hitherto held the nuclear family together. This elemental social unit seems to be being drawn apart by an irresistible centrifugal force.

**Domesticating technology**  How far will this centrifugal process go? It's difficult to say. Certainly, in an important respect, it's a process of emancipation. The range of choices available to the individual is so much broader than it was only decades ago.

Such a process will not be reversed: that much is sure. On the other hand, it's a truism that human beings are essentially social animals. We'll continue to live together in homes. The way of the hermit will be, as it always has been, only for the few. Social contacts will continue to be important in our lives. But they'll be contacts which are genuinely of our own choosing. We'll choose them from a wider range than ever before. And we'll experience them in a greater variety of ways and through different channels, many of them electronic.

As electronic media come to offer increasingly richer experience, physical absence will be less of a barrier to satisfying social contact. We'll see and talk to each other through videophone, we'll correspond instantly via electronic mail, and we'll party using videoconferencing. We'll still enjoy the physical presence of our friends as we do now, but, in addition, we'll 'entertain' guests who are thousands of miles away.

But amplifying the individual's social reach in this way is not the only manner in which technology can benefit the residents of the polycentric home. Communications with the world outside – where the word 'world' should be taken

literally – will also give people access to many interactive sources of information and knowledge, amplifying people's cultural and intellectual experience. A whole gamut of new tele-services which can be used and enjoyed remotely will enter the home in the same way, from traditional consumer services such as shopping and banking, to video-on-demand and other sources of entertainment. And continuing advances in sound and image processing and transmission will inevitably amplify still further our sensory experience. Perhaps faster than we can now imagine, our experience of sensory stimuli which originate at another time and place will approach the quality of 'live' experience.

Technology can fulfil many of the desires people have at the end of this twentieth century in helping them live a varied and wide-ranging life. But if it's not to undermine our sense of stability and belonging, if we're not to find ourselves adrift in cyberspace, lost and unable to call home because we don't know where home is, then we must make sure now that technology becomes properly integrated into the home, that it becomes, as it were, domesticated, serving and not threatening its masters.

**Designing reconciliation**   I see it as the task of industry to suggest ways in which people can both enjoy the tremendous benefits technology offers and at the same time enhance their traditional values and experiences. It will be up to designers and manufacturers to give shape to the myriad new interrelations which will arise between people, objects, space, time and circumstances, to bring together past, present and future in a home environment which is both highly stimulating and deeply satisfying.

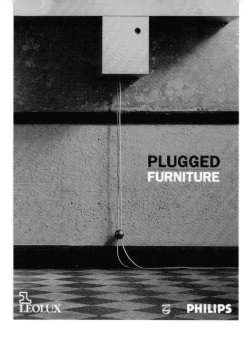

*Plugged Furniture*: integrated media furniture by Philips and Leolux – *Ironie*, *Tavoli* and *Parete*

The challenge, in short, is to reconcile technology and traditional domestic qualities, to make new objects and new media at home within old walls.

**New Objects, New Media, Old Walls**  In 1995, under the title *New Objects, New Media, Old Walls*, we embarked on a project to explore ways of reconciling technology and traditional domestic qualities. The aim, as in the case of our earlier project *Television at the Crossroads*, was to gain feedback from around the world by taking an exhibition with the tangible results of the project on tour. Extrapolating from already existing or incipient technologies (though in some respects relying on no more than imagination), we envisaged a home in which present and future technologies could be integrated in an essentially 'human' way. The result would be a home that, despite its advanced infrastructure, felt comfortable and full of traditional domestic values. We assumed that it would have its own internal communication system, processing information derived from external and internal sources. Input from the outside world would enter through a telecom 'gateway', ushering in all kinds of digital services. Output generated by the residents would exit via the same route. The family's own collection of CDs bearing favourite music, video, games, educational and informative material would be stored in a central archive.

The system, linking up a wide variety of objects, was assumed to be cordless. This would allow objects to 'be themselves': that is, like those of a pre-electric age, they would be locally independent of their power source and therefore easily movable. Power would be provided by batteries which are recharged in a way which is least burdensome. We assumed that the object merely had to be placed on a special small table mat, and the batteries would be recharged. The system would be accessed using intelligent, versatile control devices operated via a touch-sensitive screen or by voice activation.

**Familiar forms, instinctual functions**  We 'furnished' the house with a variety of new media products that, although not yet even in the pipeline, could conceivably be produced within the next five to ten years. To help the imaginary users feel – literally – at home, we gave many of the objects familiar forms. The CD archive unit resembled a chest of drawers, the traditional archetype of a storage device. A remote-control access screen was reminiscent of a framed photograph standing on a table. Another control device looked, felt and functioned like a leather-bound book. Loudspeakers were no longer square boxes but had the form of vases or urns. Terracotta dishes in the kitchen turned out to be display screens.

The objects were not only integrated into the domestic setting in terms of their visual and tactile characteristics. Their very nature was designed to respond to deeper psychological needs. The thrill of collecting things, with its origin in our primitive hunting past, and the pride of displaying our trophies and possessions, are deep-seated human instincts. As digitalisation and miniaturisation expand their domains of application, collections of books, records, videos, and the like, will, in visual terms, be diminishingly impressive. To compensate for this 'impoverishment' of our instinctual envi-

ronment, the Archive Unit was designed to be capable of being expanded by adding extra modules as the user's collection grows, reflecting the presence of more data in the system. In the same way, we took account of the traditional, reflective mode of browsing to make a selection in the CD Album, which contained the booklets that accompany the discs. And just as we can start reading when we find an interesting page in a book, so we could activate the data carried by the CD of our choice by touching the appropriate page in the Album.

**Plugged Furniture** *New Objects, New Media, Old Walls* looks perhaps a decade ahead. But we were impatient to get moving: what steps could we take immediately? Through that project, we came into contact with others who, within their own fields, were also seeking to address the same sorts of issues but from other directions. Furniture designers and interior decorators, for example, are regularly confronted by the discrepancy between the living environment and technology, and it was therefore only natural that we and Leolux, the Dutch furniture design company, should 'find each other'. Philips and Leolux work in complementary markets, and a symbiosis of their strengths promised the possibility of introducing the public today in a very real way to the new dimensions we had envisaged in *New Objects, New Media, Old Walls*. The result of our collaboration was a range we called *Plugged Furniture.*

**The cultural assimilation of technology** To determine the appropriate level of cultural assimilation for the video and audio technologies we planned to integrate into the furniture, we traced the evolutionary pattern followed by earlier technologies. Initially, new technologies enter the home in a relatively 'bare' form. At that stage, the physical appearance of the device itself is of no importance to us: it is in itself 'magical' enough. But, as time goes on, that often quite utilitarian look acquires a special appeal in itself: it signifies the presence of something highly valued – the latest technological miracle. However, as the technology itself becomes commonplace, our attention shifts towards the content, that is, the quality of its functioning, and it no longer makes sense to demonstratively announce the fact that we possess such technology through the medium of a 'high-tech' exterior.

**Ironic totems** At that stage, two things may happen. One of these is to give the object an appearance which conveys some other message, some other – totemic – value. This new 'meaning' may be almost anything: a witty statement, a stylistic reference, or an icon of some lifestyle from the present or past. Such forms generate curiosity and interest, not now for the technology, but for the more subtle, often ironic message they embody. They combine the new and the old, not denying either (by pretending that the object is actually old or that the latest technology is not present), but juxtaposing them in a playful statement of the owner's contemporary values.

It was this evolutionary path that we explored with *Ironie*. This comprises an audio unit and a TV/VCR unit. The irony lies in the fact that their forms are inversions of each other. The audio unit is a sturdy shelf with fixed bookends at each end and a small cupboard centrally positioned beneath it. The bookends contain the loudspeakers and the cupboard conceals the audio equipment. In the case of the TV/VCR, the 'feet' (corresponding to the bookends of the audio unit) provide a surface for the VCR cabinet and a cushion. The element corresponding to the cupboard of the audio unit is the TV.

**Disappearing act**  An alternative to giving the object a totemic appearance is to have it more or less disappear. Since a 'high-tech' appearance has been deprived of its special significance by the widespread adoption of the technology, the object can now be hidden away or (in the context of current miniaturisation technologies) be reduced in size and prominence. One strategy for reducing the prominence of a technological device is to integrate it with another object. This may involve combining two or more similar devices to save space (for example, the tuner-amplifier or the combi-TV-video). Alternatively, a technological device can be integrated into a non-technological object, such as a table, a chair or some other item of furniture.

Taking this approach, we came up with *Tavoli*, a set of units which can either stand separately or be placed together to form an elegant table or sideboard. Each unit contains an item of audio or video equipment – a television, a VCR with storage space for cassettes and CDs, audio equipment or one of the loudspeakers.

**Saving space**  Finally, we looked at a further development of the above strategy, namely to integrate the objects into part of the more permanent infrastructure of the domestic space, such as the walls or ceiling. As in the case of *Tavoli*, such integration saves space and is aesthetically attractive because it reduces the multiplicity of individual objects around the room. With society becoming more individualised and as we surround ourselves with more and more technological devices, this is a benefit that is becoming highly important. The question of domestic space – and particularly how it can be used most efficiently and flexibly – is likely to become urgent in the near future.

The polycentric nature of the modern home was recognised as long ago as the 1960s, with the development of open-plan architecture for homes. But although this approach acknowledged that rooms were no longer monofunctional, it failed to foresee the individualisation of society, which would mean that a number of separate – though now polyfunctional – areas would still be required. Today, there seems to be a need for movable or 'virtual' walls. Open-plan architecture similarly failed to provide an answer to another problem that has resulted from the rise of the polycentric home. Since people nowadays want to watch television not only in the main living room, but in the kitchen, the bedroom and even in the bathroom, we find ourselves having to duplicate that technology in various parts of the house. The same applies to radios and audio systems, and, increasingly, to computers. This duplication in terms of cost and space is far from ideal. We need to try and develop a new flexibility, not only in the segmentation of the living space, but also in the manoeuvrability and portability of objects, such as televisions.

Taking this as our point of departure, we developed *Parete*. This forms a dividing wall between two areas. It incorporates TV and audio equipment in open cupboards in such a way that they can rotate through 300° to face areas on either side of the wall, as required. Shelves above each cupboard can accommodate books or vases, and cassettes and CDs can be stored in a vertical drawer in one of the narrow sides.

**Coming home**  Despite the radical transformations that are affecting the way we live, our deeper nature remains the same. Like our ancestors, we need somewhere we feel secure, somewhere we can rest, where we belong. Amidst tumultuous change, when the future (and even the present) seem so uncertain, we need – perhaps already even long for – somewhere to come home to.

The work I've described here is a first attempt to respond to this growing *cri de cœur*. And given the accelerating pace of technological developments, further exploration of this dirction is a matter of some urgency. If 'ordinary people' (and we count ourselves among them) are to accept and maintain control over the technology that will enter their lives – and if it's to benefit them as much I believe it can – then it's clear that the shift from hardware to humanware cannot be delayed. Public debate on the issues involved is essential, because I'm convinced that it is only through dialogue between designers, manufacturers and consumers that we'll be able – together – to design the reconciliation of high technology and the human tradition.

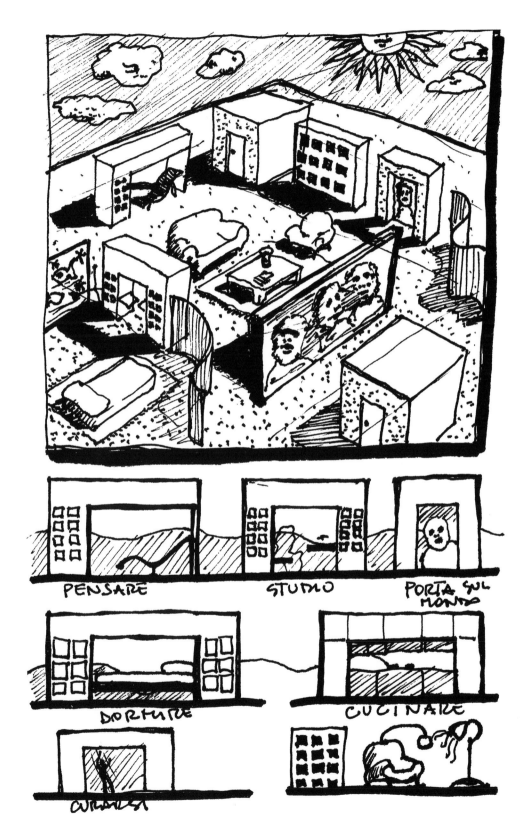

PENSARE

STUDIO

PORTA SUL MONDO

DORMIRE

CUCINARE

CURIOSITÀ

# A Design Recipe for These Times

October 1991. It was one of those bright crisp mornings you get in Northern Italy in early autumn. The sun was glistening on the still waters of Lake Orta, the purple silhouette of the Alps in the distance. That we should be meeting in the heart of Piedmont, with its noble and long tradition of gastronomy and culinary art, was very fitting. For what we – Alberto Alessi of the Alessi design company, Kees Bruinstroop of Philips' Domestic Appliances division, and I – had come to discuss over lunch at a small lakeside restaurant was how, together, we might challenge the prevailing ethos of the western kitchen. We wanted to explore the possibility of forging a link between what to an outsider might seem to be complete opposites: the small but influential Italian firm of Alessi on the one hand, and the vast multinational Philips on the other. But all three of us felt that the time was ripe for an exciting new development. Each of the two companies had something the other lacked. A productive collaboration would not only be beneficial to them but to the wider world as well. Over the local speciality, *risotto con tartuffi bianchi*, we talked and planned.

**Where did they go?** The earlier months of that year had been very eventful for me. After several years of working in Italy, I was about to move to Holland to join Philips as head of Corporate Design. I was keen to put into effect my thinking on what I felt needed to be done in the area of consumer electrical goods. I'd become convinced that the spirit of the industrial revolution, based on the maximisation of quantity at all costs, had resulted – in spite of all the benefits it had brought – in many of the old cultural and essentially human values being lost from our lives. The result: alienation and emptiness. No single company or nation has been to blame; it's as if the very nature of our society made it inevitable.

Once, the relationship between user and product was one of trust and affection: the tool which served us well, the shoes which were so comfortable, the heirloom which bore witness to our history. But almost within living memory, the throw-away society – a world which prizes newness above proven worth, and flashy gimmickry above genuine usefulness – has lured us away from the values of the past. We've forgotten how to care for objects, forgotten that they are, as Ezio Manzini of the Domus Academy in Milan put it, "creatures produced by our spiritual sensibilities and by our practical abilities", which, once created, have lives of their own, needing us as much as we need them. And it's not only the world of objects we've lost touch with. In an exactly parallel way, we've become distant from the natural world. Only now, as we face the possibility of global catastrophe, are we beginning to appreciate that the respect and reverence which our distant forebears showed their environment was actually a reflection of wisdom greater than we, with all our scientific knowledge, have managed to muster.

What we need now, therefore, is a radical transformation of our points of reference, the values and criteria we've been using to evaluate the relationship between ourselves and our environment, between us and our objects. And this entails a profound change in the culture of design.

**Rehumanising the kitchen** Fortunately, in certain areas of life, we're beginning to see a reaction against this dehumanisation. Pure techno-fix solutions are becoming less attractive. And one of the places this renaissance is happening is, appropriately enough, what used to be the heart and soul of home life – the kitchen. A revaluation of tradition, creativity and ritual in that central family space is in progress; and there's evidence of a renewed concern with nutrition, flavour and the freshness of foodstuffs.

But why did we come to lose all this in the first place? For the very good reason that women in our modern society rightly refused to devote most of their lives to slaving over a hot stove. The trend of the past century has been towards spending as little time as possible in the kitchen. The recent development of the kitchen environment has followed Taylorian time-and-motion lines; and under the influence of home economist Christine Fredericks, the kitchen became the embodiment of efficiency, cleanliness, and tidiness. Surfaces and appliances had above all to be hygienic and easy to clean. Speed of food preparation became the watchword. The more mechanical, robotic and controlled, the better. In the futurist ideal, the kitchen was fully automated, a place where one need not appear or, if that were unavoidable, a place from which one could quickly escape again. It relegated food to the level of the vitamin-and-protein pill. Why should we waste our time with all that bothersome cooking?

It was this philosophy that generated the white plastic utensils and appliances that have become so familiar: the kitchen as impersonal food factory, nutrition clinic or chemical laboratory. Emotion, creativity and indeed even flavour were banished. What happened to the warm country kitchen, filled with delectable smells? What happened to the triumphant look of pride on the cook's face as the masterpiece is placed on the table amid the ritual of the feast? What became of what Elizabeth David once called "the most comforting and comfortable room in the house"?

Life in these latter days of the twentieth century is an accelerating rush; it's unlikely ever to slow down. All the more need then to find time for moments of calmness and serenity; to reassert the value of *being* over the virtue of *doing*; to substitute in due season the joys of creation and socialising for the strains of competition and endeavour. What better place for this than the kitchen, and what better time than mealtimes?

But we can't just turn black the clock. We've become used to modern conveniences; our lives are fuller and more varied than those of our ancestors, and we don't want to spend all day bent over the oven like the tireless *bonne femme* of previous generations. What we need to do is rediscover the virtues of the traditional kitchen without reshouldering the burdens it imposed. We need to find modern kitchen tools which are more congruent with the age-old culture of cuisine, gastronomy and mealtime socialisation; equally as effective as our current appliances but without the noise and limited manoeuvrability their electric motors entail.

Above: Integrating cultural icons into table-top cooking: sketches for product concepts

Right: Stefano Marzano with Alberto Alessi

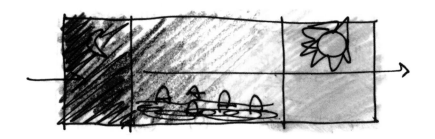

**Poetic tools**   It was ideas such as these that were occupying my mind when I was appointed to my present post at Philips. As luck would have it, in that curious period between the actual appointment and its being made public, I was conducting a graduate seminar in Design Management at the Domus Academy. As a case study, we'd chosen to look closely at the Alessi company. Now, suddenly, with the secret knowledge that I was about to start work for one of the major multinational producers of electrical kitchen equipment, I began to look at Alessi in a rather less academic and more practical light. I probed, I questioned, I listened. And the more I learned about the company and its way of working, the more I realised that if only something of their culture could be introduced to the mass market, it could exert an enormous influence for good on the lives of many people around the world. It could help bring about the profound change in design that I saw as urgently necessary.

Alessi are clearly expert in designing tools that are simultaneously works of art, items to be cherished both for what they do and for what they are. Talking with the Alessi brothers and their craftsmen at their workshops in the Piedmontese hills, I sensed an atmosphere of what I can only describe as a truly Renaissance devotion to beauty, expressed in the creation of objects of everyday life.

As befits such a scene, perhaps, there was little evidence of high tech around, and certainly not in the products themselves. Of course, mass market kitchen appliances without high tech are virtually unthinkable these days. And it is expertise in precisely this area that Philips could supply. The more I thought about it, the more obvious it seemed: we had to work together. Philips, a world leader in kitchen appliances, experts in miniaturised electrical and electronic technology, skilled in designing ergonomic products with a wide variety of functions, possessing knowledge of the needs of a broad public and operating a worldwide marketing infrastructure; and Alessi, begetters of aesthetic and creative insights, heirs to a long tradition of individual craftsmanship... The synergy that could result from combining the strengths of the two companies in some way was an exciting prospect. Instead of the sterile, cold kitchen gadgets which had been typical of the second half of the twentieth century, we could produce warm, poetic tools with no loss of functionality and with an enhanced humanistic character. Tools that would help recreate traditional kitchen values for the twenty-first century.

**Serendipity**   I longed for the opportunity to talk about this exciting possibility with someone openly. Again, serendipity played a key role. Kees Bruinstroop, whom I knew well, happened recently to have become head of Domestic Appliances at Philips. As old friends, we met in July 1991 to catch up on each other's news. Kees told me of how he was planning to meet his new challenge; and I shared my recent thinking with him, how I felt the time was ripe for a shift from hard, technological values to softer, more human ones, and how I thought this could be reflected in the kitchen environment. I mentioned my encounter with Alessi and the train of thought it had triggered in my brain. We soon discovered we were operating on the same wave-length, and by the end of the afternoon, we had agreed to pursue further the possibility of a joint project between Philips and Alessi. It was too good a chance to miss – why wait? I phoned Alberto Alessi the next morning. The holiday season was upon us, and as we had all planned to go away at different times, we couldn't meet straightaway. But we arranged to get together in Piedmont to discuss possibilities at the earliest convenient moment, in October.

It's a commonplace observation that as one gets older, summers get shorter. Those endless summer holidays of childhood never come round again. But although it wasn't endless and despite all I had to do, I know that summer of '91 passed more slowly than any I remember as a child. Finally, October came. We met as planned, and, as the cliché goes, the rest is history. The Philips-Alessi line products were launched on 27 September 1994.

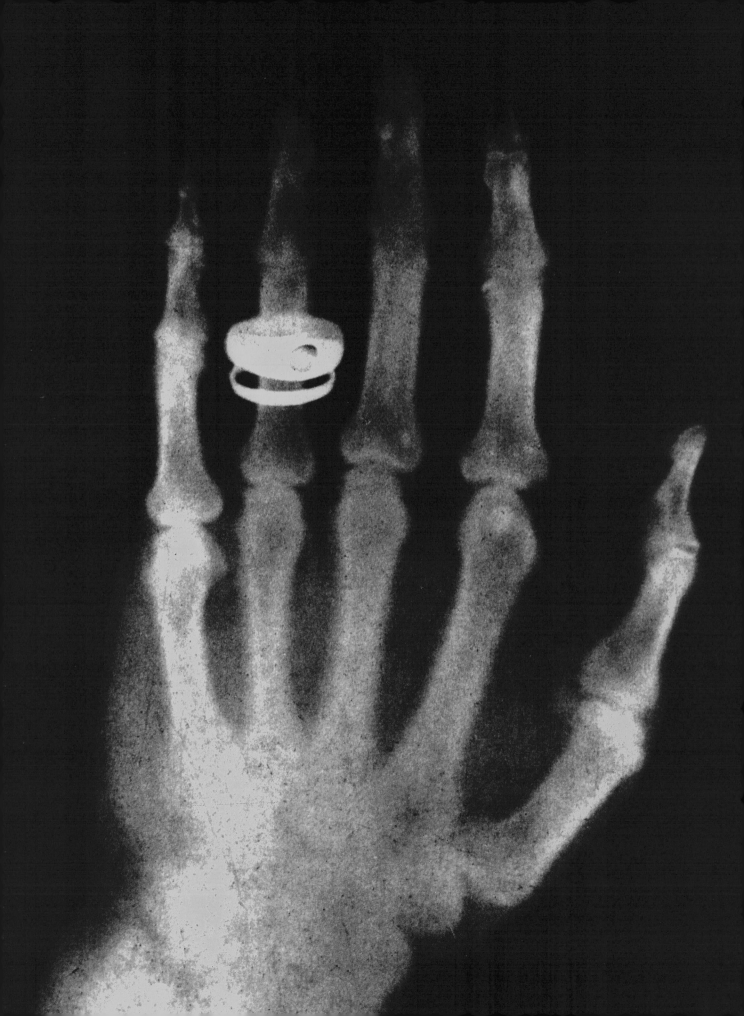

# Humanware in Healthcare

**Constant throughout history** Whenever we see pain, suffering, fear or despair in a fellow human being – however much we may have become hardened, however far away the sufferer may be – we instinctively respond as humans have done throughout history, with a primal urge to help, to try to relieve suffering by giving comfort, consolation, protection.

In our society, those who express this humanity most concretely and effectively are the medical community. This was the case in the past just as it is today. Throughout history, this instinctive reaction to help has inspired creativity and inventiveness – and it continues to do so.

Despite this universal, timeless and quintessentially human basis, healthcare systems today are increasingly confronted, like so many other areas of life, with hard economic realities. Within little more than half a century, medical practitioners have progressed from being seen as brave fighters in an often unwinnable battle, to their status today as miracle workers for whom almost nothing is impossible. The discovery of antibiotics, genetic engineering, and amazing technologies which amplify our senses of sight and hearing, such as x-rays, ultra-sound and magnetic resonance imaging, have developed the expectation in ordinary people that there'll always be more to come and that they have an absolute right to optimal healthcare, whatever the cost.

Unfortunately, however, as we all know, realism demands that the healthcare system is run not only with maximum effectiveness, but also with maximum efficiency. Today's challenge is therefore to find a way of reconciling the timeless human impulse to help another in pain or suffering – whatever the cost – with today's rigorous economic constraints. But as healthcare administrators struggle to balance medical ideals with practical economics, it's important that patients and doctors don't lose sight of each other, that we don't end up in a Brave New World in which the altruistic motives which have guided the medical profession throughout its history become lost and in which patients feel abandoned by their human healers and left to the mercy of machines.

**Designing medical equipment** Let's take a closer look at the specific problems involved in designing medical equipment. Normally, when we're designing an ordinary electrical product for use in the home, the buyer is also the person who's eventually going to use the product. But in the case of medical equipment, hospital managers will buy the machine, doctors and nurses will operate it and, of course, the patient will be on the receiving end. That means we have *three* groups of people to satisfy, not just a single individual or family. And these three groups of users have different priorities and concerns. Let's take a look at the sorts of factors we as designers have to take into account.

**The hospital environment**    First of all, the hospital environment itself – by which I mean the physical environment. Obviously, any product that comes into a hospital must be able to be kept sterile. It must be easy to clean and able to withstand any aggressive liquids which may be used. In many hospitals, space is at a premium. This means that equipment, which is often quite large, must take account of the size of doors and windows and the height of ceilings; and, of course, the general dimensions of the areas in which they are likely to be used. There are, I suspect, few operating theatre staff who can perform at their best with a colleague's elbow sticking in their ribs. And we should not forget that unsung hero, the service engineer: he also needs room to work. We often say 'Time is of the essence', but sometimes space is every bit as important in allowing people to do their job well. A third aspect of the hospital environment that designers have to bear in mind is the traditional hospital colour scheme of white or light colours. For very good reasons, I believe, the medical world is not about to ask for a scarlet scanner or a bright green x-ray machine. But whatever the basis of the traditional colour scheme, it's a limitation designers have to take into account.

**The medical team**    Let's turn now to the medical team. What are their particular requirements that we need to meet? Well, first of all, perhaps, the controls and the computer screens displaying data need to be clear. And with certain equipment, we need to take account of the fact that radiation may mean that staff have to be able to operate the controls while wearing protective gloves. Another problem is eye-strain caused by difficult lighting conditions – at one moment you need very bright light to be able to operate, while at another you need subdued light to be able to look at a screen. The effort of continually re-adjusting from one to the other can be very tiring on the eyes.

**The patient**    Practical considerations we need to take into account when designing for the patient include comfortable, smooth movement, for example. Being inserted into a big machine is worrying enough without the mechanism being jerky, noisy and uncomfortable. And since movement is involved, it's also important, obviously, that nothing can get tangled up or caught in the mechanism. Equally obviously, protection from unnecessary radiation is also a key consideration.

Medical equipment needs to be designed with the
interests of patients, medical staff and administrators
in mind

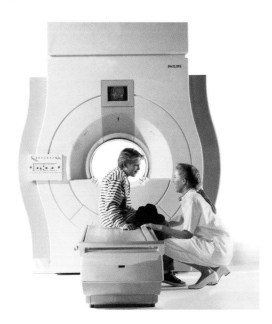

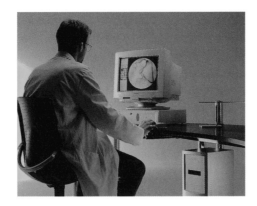

**Psychological reactions**  The above are all very practical things that designers have to keep in mind when designing medical systems. But in fact the more psychological and emotional reactions of the three groups of users are also extremely important.

We try to design our products so that they trigger particular reactions from people. From hospital managers, for example, we want to hear things like "This is our best buy in years" – in other words, that it's cost-effective. We want to hear that it's easy (and therefore cheap) to install, maintain and keep clean, and that it stays looking good for a long time.

From the medical team we aim for reactions like "I hardly notice it's there" – in other words, that it's effective in a quiet and unobtrusive way, allowing the team to concentrate all their attention on the patient without being distracted by the technology. We also try to give the product an intuitive mode of operation, one that doctors and nurses find natural. This means that the equipment almost becomes one of the team, working together with them in harmony, rather than having a mind of its own. We want the medical team to be surprised at the speed at which the equipment performs. That's not only good for the patient, by being less intrusive, but it's also good for staff, giving them more time for training and education. We're also looking for reactions like "I could do this all day", indicating that the clarity and sensitivity of its controls makes the equipment not tiring to use. And we take spatial and aesthetic aspects into consideration so that staff feel that their working environment is comfortable and pleasant.

From patients, we like to hear reactions like "I'm in safe hands here, everything's all right." In other words, we want them to feel reassured and confident in what can often be a rather frightening situation. And we want the patients, like the staff, to be surprised at the equipment's speed of operation and its comfort.

**Technologies**  Achieving some of these reactions will depend to some extent on technological developments, involving the application and refinement of skills already being developed in non-medical fields. Before long, a lot of hospital treatment and administration will be screen-based. Converging technologies will also give rise to networks of medical practitioners of various sorts, linked by satellite over vast distances.

Although these developments are basically technology-led, they have enormous implications for design in the medical world. On the most immediately obvious level, they will present doctors, nurses and administrators with vast amounts of data in visual form. If this data is to be useful and not confusing, it must be structured clearly and in a way which makes it easy to understand and assimilate. This means that software engineers and designers need to work closely together with experts from the field to work out the best ways of presenting and structuring this information.

In the second place, these upcoming developments may change the nature of the hospital as a care centre. Surgical robots combined with satellite-linked telecommunication networks will mean that a surgeon will not actually have to be present in the same place with the patient, but may be miles away. They will be able to operate on people in other countries or remote districts, or help paramedics assist casualties in ambulances long before the ambulance arrives at the hospital. All this may require some radical rethinking in the design of hospital environments and medical systems.

Another development which is just ahead is in the area of controls. Many people find the controls of high-tech equipment too complex. They have difficulty in programming a video recorder or transferring a phone call to another extension. For those people, help is at hand. A great deal of work is being done at the moment on the development of more natural ways of communicating with machines. Many products will soon be activated by talking to them in a simple, straightforward way. In the hospital environment, voice activation should make the operation of equipment during surgery, or in situations where the operator has to wear protective gloves, much easier than it is at the moment.

The technologies for voice activation and voice recognition are becoming increasingly effective. But we must make sure – and this is part of the designer's job – that the sort of thing the machine needs to hear will be the sort of thing we feel is *natural* to say. Conversation with other human beings is so natural to us, that it never occurs to us that there are certain rules and structures involved, because both we and the person we're talking to automatically follow them. When we talk with a machine, however, only one of us knows the rules naturally, and the other has to be taught. In other words, the conversations have to consciously designed, just as, today, the interaction between a user in terms of buttons, lights and screen information has to be designed.

One further development which will increasingly affect the medical field is miniaturisation. It should soon be possible to produce a miniaturised scanner, for example. This miniaturisation will obviously benefit hospital settings where at the moment certain equipment tends to take up a lot of space, is difficult to move – and is rather frightening for patients.

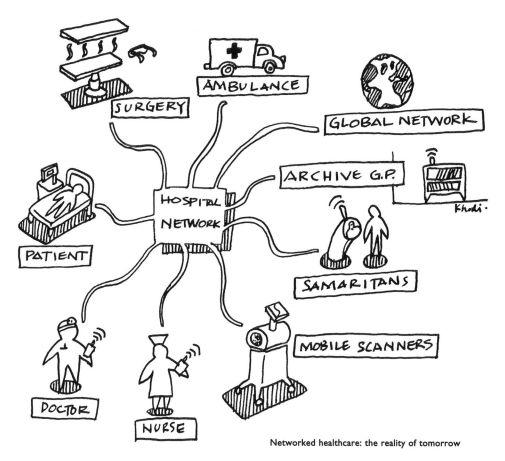

Networked healthcare: the reality of tomorrow

 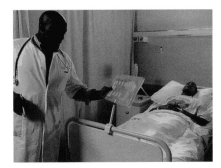 

**The human touch**  This brings me back to the most important people of all – the patients, the *raison d'être* of medical technology. Let's consider the messages that the physical shape and colour of a product convey to people, the subtle messages which determine how they feel about it. The human mind makes all sorts of associations between form, colour, smell and so on, which go right to the depths of our being. I'm sure many have had experiences like that related by Swann at the beginning of Marcel Proust's great novel sequence *A la recherche du temps perdu*. The whole atmosphere of his childhood, together with all its emotions of warmth and security, came back again in an instant when he tasted a piece of madeleine cake soaked in tea, a taste he had often experienced as a child but had quite forgotten until that moment. As designers, we try to make use of some of these deeper, pleasant associations to help people feel better in what is often a stressful situation.

These associations form, in fact, a sort of language, a universal language we all share. What we try to convey with it is, if you like, the sensory equivalent of those kindly words of comfort and consolation which turn pain and anxiety into comfort and reassurance. We want to offer consolation, to remove any distress a patient may be feeling. We want the patient to sense an atmosphere of protection, shelter, kindness, sympathy and caring. In other words, we want to provide as human an environment as possible. The sorts of elements or vocabulary we use are sensory experiences which avoid extremes, which calm and relax. Rounded shapes, soft surfaces, quiet sounds, subdued lights, gentle movements, pastel colours, temperatures which are neither hot nor cold. And we try to link such sensory 'words', if we may call them that, together to form sensory 'sentences' – combinations of shapes, colours and impressions which together convey the emotions we would like the patient to feel. These sensory sentences often have a primal, human form. The warm embrace of arms, a comforting hug; gentle hands, stroking and caressing, soothing away anxiety; a friendly arm around the shoulder; caring eyes, a reassuring smile; a mother's breast, the security of the womb.

**Partnership**  And here we come back full circle to the quintessentially human emotions and response I spoke of earlier. Medical technology is advancing at a tremendously exciting pace. In partnership with technologists, the medical community and potential patients, we, at Philips Design, are doing our very best to ensure that the humanity which has always inspired the medical profession will not be lost behind the machines, and that patients, doctors and nurses will all feel they're surrounded not by more and more technological hardware, but by understandable and friendly humanware.

Leading the way for doctors for over two and a half thousand years, Hippocrates promised he would prescribe treatment for the good of his patients according to his ability and his judgement and never do harm to anyone. I like to feel that technology is helping today's doctors fulfil this timeless creed to a degree that would have astounded and thrilled Hippocrates: increasing their skill by allowing them to give more precise and less invasive treatment; expanding their ability by allowing them to look deeper into the human body and far beyond their immediate location; refining their judgement by giving them access to more relevant information about the patient and to advice from colleagues around the world; and finally, enabling doctors to do this in a way that properly reflects the full humanity of that very special relation between practitioner and patient.

# New Frontiers in Lighting Design

**Technology and design**  It seems to be a characteristic of humans not to want to pass up any opportunity for expressing their personalities, ideas, or values. The wide variety of forms taken by lamps, luminaires and other lighting products is a prime example of this. Most sources of artificial light today are embedded in some sort of carrier. This is designed to enhance the quality of the light in at least two ways. In the first place, it modifies the colour, shape, direction and distribution of the light; and in the second, it contributes aesthetically to the larger setting, either complementing the architecture or forming a striking feature in its own right. In both respects, the carrier increases the sense of comfort and pleasure we derive from the light itself.

Philips is involved in raising the quality of our products in a number of different ways, including customer friendliness, environmental care and energy conservation. And technological innovations are set to take these improvements even further. But how will such developments affect the *design* of lighting products? I believe they'll open up new vistas for designers and architects to make people's experience of light more personal, pleasurable and profound.

**Miniaturisation**  Over the past decades, many everyday products have been miniaturised beyond recognition. This process is now affecting lighting technology. Miniaturisation is freeing designers to create products in which the light source can be made as visible or as invisible as desired. Professional systems can be made less intrusive; and in the home, more than ever before, the light source can be embedded in an aesthetic form expressing the customer's own cultural values. Lighting in the urban setting, in streets and public places, will not only be able to provide a greater sense of security, through improved directionality, dispersion and intensity, but it will also contribute to the aesthetic appeal of the surrounding architecture, by its unobtrusiveness, for instance, or by its versatile incorporation into appropriate street furniture.

**Natural lighting**  The rhythm of night and day clearly affects our physiology and our behaviour. In this respect we're like virtually all other creatures on this planet in being conditioned by our environment. But as human beings, we've also developed cultures and social structures. We give meaning and symbolic value to significant elements in our environment. One of those elements is, of course, light. We need to take this meaning, this value, into account when considering the prospects for the future of artificial lighting. It is, I suggest, one which people consider to be integral to what they see as 'natural' light. It's also an aspect of light which we can recreate and offer to the consumer as a means of improving the emotional and expressive sides of their lives, giving artificial light a new cultural dimension.

This expansion in the possibilities for cultural expression shows that electrical lighting technology has truly come of age. New technologies are usually introduced to the public in a pure, unadorned form – the technology itself is intriguing enough. Only later, when it becomes part of everyday life, does it become a vehicle for expressing cultural stratification. A recent example of this process is the energy-saving PL lamp. As it has gained in popularity, so the variety of compatible fittings has multiplied. As users, we choose lamps not only to suit our lighting purposes, but also to satisfy aesthetic and cultural needs. And as miniaturisation continues and technology advances, the opportunities for expressing personal values in lighting products will grow.

**Sun, moon and stars** From the very earliest times, it seems, we've bestowed special significance on the sun, as the source of light, warmth, safety, and ultimately life. The light it produces has been identified, either literally or metaphorically, with the godhead in all the major religions. Lighting and fire, whether permanent or occasional, generally signify a sacred or spiritual presence, an offering, prayer, intercession, or purification. They're often viewed as sacred or even of divine origin, and sometimes even directly identified with the deity.

By association, light has also come to represent understanding (we call it also 'enlightenment', and 'illumination'). We see it as an ultimate goal. All around the world, festivals of light are held, at which artificial light (torches, fires or candles) is used to honour, represent – or even be – the deity, forming an essential part of some of the spiritually most important rituals we perform. By contrast, the absence of light has been associated with evil, danger and death. Indeed, of course, for our ancestors, dependent on their sense of sight to guarantee their physical safety from predators, absence of light was probably very often frightening and perilous.

On the other hand, the night, when dimly illuminated by the moon and stars, also gained other associations: those of magic, mystery, awe and intimacy. By reducing the power of vision, darkness creates uncertainty, generating questions, arousing curiosity. At the same time, it offers privacy, to be enjoyed alone beneath the enormity of the night sky, or with another human being, equally mysterious and awesome...in another sense.

Traditionally, too, the rising and the setting of the sun have aroused special emotions in human beings. The increasing light of day stirs daytime creatures into activity and bustle; and as light fades, noise and movement decrease, in a continuing pattern of crescendo and decrescendo. The changing intensity of the light and its varying colours together with the sense of developing activity or increasing repose – all these associations contribute to the fascination of these particular moments in the daily solar cycle. In regions farther away from the equator, the longer periods of twi-

light exert an extremely powerful effect on the emotions, calling up memories, encouraging longings – taking us out of the present into times that are normally beyond our reach.

Twilight has been exploited by many artists to generate poetic or emotional responses. The seventeenth-century French painter Claude Lorrain, for instance, in his idealised, idyllic landscapes, used a soft, suffused yellow glow of approaching day to suggest a golden age. The romantic German painter Caspar David Friedrich also often depicted such key moments, giving them, in the case of sunsets or rising moons the significance of approaching old age and death, linking the cycles of human life with the cycle of night and day.

**Light and shade**  Not only the intensity and colour of light in the sky has important cultural and aesthetic significance. Since light has the property of casting shadows, we also need to consider the effect of light and shade in particular geographical locations and in relation to natural or artificial objects, such trees or buildings.

**Geographical location**  Light and shade are not the same in all parts of a country or in all parts of the world. The intensity of light varies, depending on latitude and the surrounding landscape.

In cool, grey northern climates, where natural light is limited, we generally value light above shade. The opposite is true in hot, bright, desert or tropical regions. On the other hand, in the clear air of unspoiled deserts, the unhindered light means that we see everything with great clarity, but we can also see so far that we lose all sense of size, scale, and distance. In southern climates, colours are brighter and luminescent, too. It is not for nothing that Provence has attracted many artists over the years. Van Gogh went there, as he said, "to look at nature under a brighter sky". By contrast, in the foggy humidity of the western coasts of Europe and North America, where the light is mediated by mists and cloud, the distances seen and objects perceived change from day to day, sometimes from hour to hour, so that we live with a continuing sense of mystery and variety.

The landscape, too, can affect our interpretation of light. In flat landscapes, the vast proportion of sky in relation to land in our field of vision means that visitors from more hilly regions are always struck by the lightness of the view. It's not surprising that these places have produced more than their fair share of gifted landscape painters. Mountainous regions, however, produce more shadows, which engage in an intriguing and sometimes threateningly dramatic interplay as the sun moves through the sky. In other words, the way we evaluate the distribution of light may depend very much on the location in which we experience it.

**Architecture**  We need light to be able to see. That means that most buildings need to have some way of admitting natural light. But this is not just a utilitarian question. Light, as it enters a building, is also a powerful, though ephemeral vehicle of expression. Because it moves, changes character, and comes and goes with its source, light has the power to give to the inert mass of architecture the living quality of nature.

Although architects do not quite control light, they can predict its behaviour well enough to catch its movements meaningfully. They channel it through openings and gaps to fall into recesses in the shape of their buildings. They manipulate it further by adjusting the angles and textures of surfaces, making it bring forms alive by creating ever-changing contrasts between light and shade.

Because of this link between nature and art, the variation in the quality and intensity of light in different climatic

The direction, intensity and colour of light can all be used to create illusions and affect atmosphere

regions plays an important part in forming local architectural styles. In hot countries, for instance, care is taken to create cool, shadowy environments within. In our part of the world, however, we value big windows and open spaces within the building in an attempt to bring as much light into our indoor lives as we can.

But there are also differences in architecture which are not so much determined by latitude as by culture. Take, for instance, the simple, light interior designs of the traditional Japanese home. There, translucent screens are used where western architectures have preferred solid walls.

Such architectural differences and varying approaches to interior design will affect what people perceive as, say, 'cosy' or 'business-like'. In northern latitudes, sitting indoors, on a dark cold evening, beside – and in the light of – a warm fire, perhaps including a candlelit dinner, is cosy to us. No doubt, it incorporates folk memories of ancestral banquets held around an open fire in the tribal hall, safe from the elements and dangers which lurk in the dark, such as wolves and bears. By contrast, people in warmer countries tend to value evenings sitting outside their homes, conversing with the neighbours, or eating in the open air, with the food cooked over a fire. In Japan, the dark and enclosed ambience which we experience as cosy and secure, would probably be felt to be claustrophobic. The Japanese environment feels cool and unemotional to a westerner.

Architects have better control of interior light than exterior light, since they can select the position, size, and shape of its source. With glass and other transparent materials they transform even its colour and intensity and so give light a meaning independent of that which it gives to the structure itself. We see this most powerfully in gothic cathedrals, where the stained-glass windows transform the rays of the sun into a mystical diffusion that seems to descend from above like some supernatural vision.

Sometimes light can create illusions, hiding rather than clarifying. When it comes out of the darkness at us with great intensity, for instance, it seems to spread outwards from its architectural source, the window through which it comes. This illusion may be used to express meanings, so that central columns supporting a row or window may seem to disappear in the bright light and the roof above seems to float freely in the air.

**Directional light**   Often the direction from which light comes can have a significance of its own. Stage lighting, for example, makes full use of this: the spotlight picking out an individual, the backlit figure silhouetted against the brightness, the footlights casting light upwards. We see the same meanings exploited, too, by photographers: the sun shining almost into the camera, long shadows cast towards the viewer. The subtle interplay of light and shade, heavily laden with meaning, was perhaps used to best effect by the American 'film noir' directors of the forties: Hitchcock, for instance, in *The Third Man*, or Billy Wilder in *Double Indemnity*. It allowed them to create strong emotions very effectively but with relatively few resources.

The direction of lighting also plays an important part in the domestic setting. The directional source of light with the most social significance of all is perhaps the fire. 'The firelight's glow' – the warmth, security and intimacy of this setting has been embedded in our popular culture from the very earliest of times. And the way in which flames flicker and vary the fall of the light is important in two ways: it varies and softens the way we see each other, and it allows us to observe the light source meditatively for long periods at a time. The same principle also accounts for the particular attraction we perceive in the candle's flame.

Many great artists have made effective use of theatrical and directional lighting. In *Christ and the Woman taken in Adultery*, for instance, Rembrandt places the Woman and Christ in the spotlight in a highly dramatic setting. In his *A Scholar in a Room with a Winding Staircase,* the main figure is illuminated by light from a window; and he usus a low source of light to great effect in *The Adoration of the Shepherds*, where the light from the Christ child radiates upwards, lighting the faces of the shepherds bent over the crib.

**New frontiers**   All these elements, and no doubt many more, go to make up the cultural and emotional meaning of light. These age-old meanings and deeply rooted associations play a vital part in determining how people perceive and experience light. The challenge – the new frontier – for us is to find ways of tapping into that depth of meaning, to explore it, to understand it, and then to codify it – to write, if you like, a grammar of this cultural and social language. Its universal elements and its culture-specific elements. If we can do that, we shall then be in a position to manipulate it, to speak that language; translating it into physical and technological specifications, and creating lighting experiences for people which will be truly meaningful. It is an exciting challenge. I believe Philips is ideally placed to push this frontier forward and enhance in this way the quality of its customers' lives.

# City Treasures

**An unbalanced response**  Under the strain of traffic congestion, air pollution and inner-city crime, the urban environment in many parts of the world is in danger of disintegrating. In large measure it is our unbalanced response to city life that is to blame. Though the human race has been eager enough to seize the economic benefits city life offered, we have failed to maintain the physical elements which traditionally made us feel at home in cities and gave us a sense of communal pride.

**Rehumanising cities**  Features of the landscape, such as rivers, hills and lakes, combined with architectural monuments have always given cities their special charm, forming a vital part of what may be called our 'natural' urban environment. If civilised urban life is to be sustainable in the long term, we now need to 'rehumanise' cities by restoring the balance between purely economic concerns and due care for the conditions which people naturally find attractive.

**The meaning of light**  Light can play an important role in this. The perception of light is a sensory experience with both physical and emotional aspects. Viewed directly, bright light can be a painful glare; viewed indirectly it can create excitement, focus attention, or arouse curiosity. Natural sources of light have also given rise to associations that are now deeply rooted in the human psyche. We experience the light of the sun as security, warmth, and clarity of vision; while the light from the moon and the stars is magical and mysterious, inspiring romance, contemplation and wonder. Such cultural and emotional 'meanings' of light should guide the future development of urban lighting, since ultimately the future of a city depends on the emotional contentment of its inhabitants. Let's look at some recent examples of work in this area that are taking the idea of City Beautification through lighting to new levels of effectiveness.

**Lighting needs**  Of course, artificial light was originally intended to improve our vision, to extend people's powers of sight beyond the hours of daylight. And being able to see more clearly meant that people were better able to avoid dangers, and they could see where they were. This rational function is still a vital one in cities; but it can often involve an unsuspected degree of subtlety.

Not everyone's needs are the same. This is particularly noticeable in today's conurbations. They're multicultural environments for multidimensional people: different ethnic, social and age-groups live and work alongside each other.

In the old days, cities slept at night. New York – the city that never sleeps – was perhaps the only exception. Now, however, many cities around the world operate on a steady 24-hour-a-day basis. This has clear effects on public lighting policy. Roads, public areas, and so on, need to be well illuminated throughout the night. And even those who are

The use of light to enhance old streets may range from
directional lighting that brings out their architectural
structure to the re-creation of domestic cosiness through
'hearths' that generate heat through infra-red light

awake at night are not all engaged in the same type of activity. Some are out enjoying themselves; others are working. The needs of different age-groups may vary, too. Many advanced communities, for instance, are ageing communities. Older people today are more adventurous than in previous generations; they're not staying at home. But when they're out and about, they *do* need to be able to see where they're walking, even in recreational areas, so that they don't stumble. The old days in which street lamps were the only sort of public lighting have gone. We've developed a taste for seeing things after dark. Attractively-lit shop windows have been around for some time now; and the floodlighting of buildings for commercial or cultural reasons is also becoming more common.

**Lighting the old streets**    Again, the subtlety that is developing here is remarkable. We're moving beyond a rational motivation to an aesthetic and cultural one, so that the qualities we value during the day can also be appreciated at night. A nice example of this is the picturesque village of Rochefort-en-Terre in Brittany. A recent project developed lighting to show off its many architectural treasures to tourists at night. Small alleys that would be uninviting at night if purely rational lighting were used, become thoroughly charming when carefully placed lighting brings out the architecture of the buildings and the rough-hewn texture of their walls. Even more strikingly, the village square, which would be almost invisible at night with its original lighting, becomes a cultural treasure.

In such settings, the source of light will often be hidden from view, concealed in the ground or elsewhere. However, sometimes the luminaire itself can become part of the aesthetic experience. In Rochfort-en-Terre, old-style lamps with sand-blasted glass are used. For more modern environments, we at Philips are particularly proud of our new LightColumn lamp, which makes use of remote-source lighting to provide effective but interesting indirect street lighting.

**The natural environment**    Buildings are in large measure what makes a town attractive in people's eyes. But we should not forget the role played by the natural environment. Trees along boulevards can become a highly attractive element in the cityscape when lit well. In fact, some might say they are even more striking at night than they are during the day, when they have to compete with so many other distracting factors around us in the busy streets.

In the lamp-lit streets of the past, we tended to look down, to see where we were putting our feet. Now, when we look up at floodlit buildings and trees, we're able to take in the wealth of architectural and natural beauty around us.

But we can raise our eyes even higher. Sometimes the buildings form just one man-made element in a whole natural landscape. We've only just begun to appreciate this and reflect it in lighting practices. Take the case of a medieval

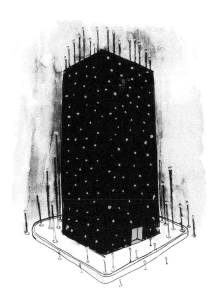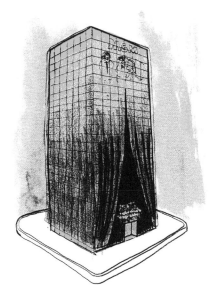

castle or fortification built on a hill overlooking a town. Floodlighting such a building is a relatively recent development in the history of lighting. Illuminating the castle without the landscape with which it interacts, however, would give the effect that it was flying in the air; intriguing perhaps, but not able to transport us back into history. Lighting the building together with its surroundings creates a total effect: depths, distance and perspective only have meaning when objects are seen in relation to each other.

**Creating new beauty**  Sometimes lighting can be used to beautify buildings which are not felt to be of particular architectural historical interest. Communications towers, for instance, often become local landmarks or orientation points, and as such, one could make a very good argument for illuminating them. They're not generally seen as part of a city's cultural heritage, but with the right lighting, they can be turned into something very special, and make an important contribution to the cultural experience of the citizens. The Retévision communications tower, which is a prominent part of the Madrid skyline, is a case in point. It has now been illuminated so that it's visible from all access routes, singled out from its surrounding buildings, which are much less interesting. Coloured light here contributes an important part of the effect. A similar project in Chile uses colour to even greater effect. The colours, which change every half hour, range from pastel to bright hues. The dynamic effect is reinforced by the movement every two minutes of vertically directed search lights.

And what if there are no interesting cliffs, mountains, castles or towers? Well, then it's possible to create such a structure with light – a virtual landscape, if you like. I recently saw in Holland a Christmas tree created by placing lights on the cables of a radio transmission tower. It provided a fascinating contrast with a nearby windmill. A similar example is that of a hotel where the roof lighting changes colour to indicate changes in the weather: passers-by can 'read' the weather as it develops.

**Bird's-eye view**  The illumination of objects like these which lie above the human-eye line is a natural extension of the original street lighting idea, which directs light to objects immediately around us. But suppose we extend that even further: suppose we consider what lighting is required when the viewer is situated high above the city or landscape. Some cities themselves lie in a valley and can be admired from the hills around: Florence, for instance. Others were founded in strategic positions on hills or rocks overlooking rivers:

Light can be used to cast an illuminated 'veil' over structures
at night or to camouflage or re-shape unattractive buildings

Edinburgh, Heidelberg and Budapest all have this sort of striking setting. The City of Luxembourg is a similar case. The medieval part of the city is perched on a rocky outcrop on a plateau overlooking the valley of the River Pétrusse. Across the valley is the new part of the city. As a result, the valley itself forms an important aspect of the city. Lighting the features within it so that they can be appreciated from above is therefore a significant combination to enhancing the city's attractiveness at night.

We shouldn't forget either that as more and more people travel by air on a regular basis, some cities may want to give thought to how their city looks to those flying in at night. We all know how passengers find city lights of any sort viewed from above a fascinating sight. Although it is not a city, but a building, the British Airways Compass Centre at Heathrow Airport in London shows the sort of thing that can be achieved. Besides an exterior that is attractive to those seeing it from the ground, the base of the building is surrounded by a continuous band of illumination which creates the impression that the building is floating on a bed of light, which is particularly dramatic when seen from the air.

Not every city would need to be lit to be seen from the air, of course. My point is simply that a city and object or landscape can often be viewed to advantage from more angles than we might normally think of.

**Worm's-eye view** Let's pursue this line of thought a bit further. We've just talked about viewing cities from the air. What about lighting *beneath* cities? As urban space becomes more scarce and expensive, many city projects may develop underground, and we may witness the rise of a new, literally '*sub*-urban' way of life. What will lighting be like in those settings, where artificial light is the norm rather than the special case? What particular needs will emerge in such an environment?

Current artificial lighting has been developed to meet human visual needs. But daylight entering our eyes not only allows us to see, it also induces a wide variety of non-visual effects. Recent pioneering research at Philips has been looking into the combination of visual and non-visual effects of lighting in interior environments. It's likely that we'll soon be able to produce biologically stimulating light, light that mimics the non-visual effects of natural light. This will make the possibility of developing communities underground much more realistic than hitherto.

We'll also need to consider the effects on people of the natural rhythms of sunrise and sunset. Recent experiments in offices have shown that, although it's now possible to regulate light indoors so that the intensity of the lighting is kept constant, people actually prefer to have the natural variations so that they can orientate themselves with respect to time in a manner which is less rational than looking at a clock.

**Communication through art** Finally, let's look at one other, exciting development. It takes lighting beyond a passive means of exploiting the beauties that exist within the urban environment; it even goes

beyond the dynamic use of light in conjunction with architecture. It seeks to combine light, landscape and architecture into a communicative medium, an international language. In the rush of our daily lives as city dwellers, we become more and more estranged from the landscape. And yet, as we know, many cultures which we tend to see as less developed, maintain a deep spiritual relationship with the landscape – to their benefit. We know how we've treated the landscape in many parts of the world, and how it might have been better if we'd had a similar respect for the environment as those supposedly primitive peoples.

The Argentinean artist, Jorge Orta, has been developing a visual, poetic alphabet, one which he calls 'planetarian', composed of universal signs and images from various cultures and ages. These form part of our collective memory and sum up the history of the human race. As the sun sets and darkness falls, they're projected onto the natural landscape. Following the natural lines and contours of the hills, they allow the landscape to speak to us.

**City treasures**  The luminographic painting of Jorge Orta is clearly one of the most strikingly original ways of taking urban and landscape lighting to a higher plane. It shows, for those who dare to imagine, that there's no limit to the positive effect lighting can exert on city life. Combining that imagination with insight into the nature of the urban experience and the nature of light itself, we can create 'natural' urban light experiences to enrich the human appeal of our cities, creating ambiences in which people not only feel safe, but in which they can regain a sense of identity and pride as members of their community. They can – once more – feel part of a shared beauty which transcends their everyday existence, and links them to each other, to their past, and to their civic destiny.

People are the life-blood of the city: their presence can be 'highlighted' by light effects responding to sensors. Here, one idea for man-sized pillars that light up in a playful, unpredictable way when passers-by trigger sensors; and another for light 'traces' that shine through a translucent surface as people walk across a square (above).

The effect of downtown landmarks that help to give cities their unique character can be extended beyond their immediate vicinity by displaying lighted images of them on the outskirts (left).

# Harmoniopolis

**The overreachers**  The past two centuries have been a time of incredible human scientific and technological achievement. But we're now gradually realising that, in a number of important respects, we've been a bit too clever for our own good. Our brains have run ahead of our nature. Nowhere is this more apparent than in the urban way of life which the majority of the world's population now 'enjoys'. Inadequate housing, traffic congestion, air pollution, an accelerating pace of life – slowly but surely, we've become estranged from our natural environment. In our cities, as in many other areas of contemporary life, it's clearly time we restored the balance between the natural and the man-made environment, between our technological achievement and our intuitive understanding.

**The natural city**  The urban settlement is a typically human product. Along with the nomadic steppe cultures, it seems to be the more or less 'natural' way we organise our societies. If we're to rediscover the sort of urban environment in which humans would feel most at home, it may be worthwhile considering what the basic elements of such early settlements were.

The earliest fixed settlements in the rich subtropical valleys of the Nile, the Tigris, the Euphrates, the Indus or the Yellow River all provided environmental factors which made living in towns relatively easy. They had climate and soil favourable to plant and animal life, an adequate water supply, readily available materials for providing shelter, and they allowed easy access to other peoples. Early on, they also developed a central place where people met, either for practical or social purposes: initially the well or watering-hole, perhaps, later the forum, the market place or the village green.

Notice how many of these features are still recognised as those which give certain cities their special charm: parks, spacious tree-lined avenues, light sociable squares, rivers and lakes, and a cosmopolitan character resulting from their convenient geographical location along major routes of communication. Add to these an interesting diversity of architecture and you have many of the ingredients of the ideal urban environment. Combining as they do the pleasures of rural life with the amenities of the city, such places seem to fulfil both our natural and our acquired needs. Contrast these with environments where such features are absent: the grim slums and shanty towns of the world's more notorious urban 'disaster areas', with dark narrow streets, no greenery, polluted air and water, traffic congestion, poverty, disease and crime...

If it's to be sustainable in the long term, any future city will have to possess the 'natural prerequisites' of the ideal urban life. It is, after all, not only unhealthy – physically and psychologically – for inhabitants to be deprived of access to the natural environment; it's also impossible to banish nature totally. The Latin poet Horace realised this long enough ago: "Drive nature out with a fork if you will: it will be back before you know it." We should therefore recognise the fact that our urban constructions are like guests in the natural environment. In the past, we've tried to impose our alien features – paving, concreting, building over the land until nature is no more than a barely tolerated guest in

its own house. Instead of the 'concrete slab' type of city which obliterates nature, we need an urban structure which leaves room for it to flourish. Urban planners have long recognised the shortcomings of the 'concrete jungle' city, though their work has often been limited by powerful financial and political interests. One of the first visionaries to point the way to combining the virtues of city life with the pleasures of the countryside was the English reformer Ebenezer Howard. Uncontrolled growth following the Industrial Revolution had led to a serious decline in the quality of urban life in England; and at the end of the nineteenth century, in an attempt to resolve this problem, Howard proposed the creation of small, independent cities which would combine the amenities of urban life with the ready access to nature typical of rural environments: he called them 'garden cities'. People, industry and agriculture were to be accommodated harmoniously within the town. At its centre there would be a garden ringed by the civic and cultural complex, including the town hall, a concert hall, museum, theatre, library and hospital. Although only two towns were built strictly in accordance with Howard's plans, his ideas influenced many later urban planners. And, indeed, they're still valid today.

**Harmoniopolis as a natural city**　How can such 'natural prerequisites' be incorporated into a modern urban environment? Urban redevelopment projects allow us to take at least a modest step into the direction of realising this objective. They provide an opportunity in microcosm to create an environment in which people can live and work in a harmonious relationship with their natural and man-made environment, interacting with it in ways that owe more to human intuition than to complex systems of learned knowledge. For this reason, we might well call such a project *Harmoniopolis* – 'City of Harmony'. Let me describe how I envisage, for example, the redevelopment of the historic industrial building in the centre of Eindhoven (the Netherlands) known as 'De Witte Dame'.

Natural light, water, flora and fauna  Within the present building, parts of the interior and roof will be removed to create shafts and atriums covering several storeys, completely or partially open to the sky, to admit natural light and air. Such areas will be a focus for natural flora and fauna – plants, trees, aquariums with fish and other aquatic life, waterfalls, and so on. The feasibility of having other types of fauna would depend on ethical considerations.

Natural light will be supplemented by artificial illumination at various levels of generality. General flood lighting will correspond to sunlight and provide an all-purpose level of light suitable for orientation. In more specific areas devoted to particular activities, directed lighting will be geared to the activity being pursued. For example, in areas in which screen-based activities are taking place, lighting will be relatively subdued; whereas for activities involving close visual scrutiny (reading books, files, etc.) a high level of illumination will be required. Wherever possible, particularly in working contexts, it will be customisable by the individual for whom it is provided.

The infrastructure (wiring for electricity, telecommunications and computers; plumbing; heating; etc.) will be accommodated beneath the floor.

Harmoniopolis will be characterised by state-of-the-art communications. 'Access to other people' these days means instantaneous global access via telecommunications, the main routes to and from the city being cables.

The Global Plaza  There will naturally be areas within the building where people can meet, or just sit and watch passers-by, like the market place, village green or town square. But in accordance with the global nature of modern society, Harmoniopolis will have a Global Plaza, a place where one can sit and relax while being (virtually) present in a social concourse in one of the world's major cities – Covent Garden in London, Times Square in New York, or Rembrandtplein in Amsterdam, for instance. A vast video-wall will show real-time pictures (with sound) of the location in question.

Architecture  An essential part of any 'natural' city seems to be streets containing a variety of buildings. Such buildings serve at least two purposes. One is the utilitarian purpose of providing a location for a given activity. The other is to express the identity of those who live or work there. Tastes differ, but the type of architecture that is most valued is usually that which, respecting its place in the whole and certain traditions, nonetheless expresses

De Witte Dame: a former factory starts a new life as

'natural' architecture

Philips Design

something unique. Buildings that exhibit originality within a coherent whole, those that make blatant or subtle use of recognisable symbols – these are the types that people seem to admire. The internal spaces of Harmoniopolis will not be the grey, characterless rooms, corridors or halls of the typical office block of the second half of the twentieth century, but individualised 'buildings', with distinctive colours, shapes and signs. From the appearance, one will be able to tell something about the activity going on inside or about the character of the occupant.

Natural signage  The iconic design of the 'buildings' in Harmoniopolis does not merely give a measure of personal fulfilment to those who occupy them, by expressing something of themselves, establishing their identity in relation to their neighbours. It also serves an important function in the signage within the whole 'city' by forming a clear landmark for those who are navigating their way from one place to another.

Animals and birds are often very skilled navigators. Migrating birds, for instance, using the stars and even man-made landmarks, manage to travel vast distances and still end up at exactly the right place. As human beings, we've developed complex orientation and navigation systems (maps and compass directions) to help us find our way from A to B. But for most of us in our daily journeys, these are too complex to use. Instead we tend to appeal to visual signals, referring to landmarks and simple orientations in relation to our bodies (straight ahead, to the right, to the left, behind, in front). This process is also simplified by the fact that some buildings have a stereotypical shape, so that, when giving directions, we can refer to their function (church, railway station, factory, office block, house) and be sure that they'll be identified correctly. In Harmoniopolis, we can maximise the possibility of using this type of 'natural' navigation by ensuring that such iconic knowledge is incorporated into the design of the 'buildings'.

Most shops these days indicate their identity with a name. In earlier, less literate times, much more iconic signs were used: someone selling boots, for instance, would hang out a sign showing a boot. In response to our increasing international (and perhaps increasingly illiterate) society in the west, we're again turning to iconic symbolism for signage in public places and for corporate identification (the logo). Apparently, an image has a more direct impact on the mind than a verbal message – a picture is, as they say, worth a thousand words. Another visual aid to navigation is colour coding, whereby colour differences indicate to visitors where they are and where they need to go.

**Even we are only human**   Humans are creatures of Nature who have in the course of time developed a complex artificial environment to answer their complex social needs. As those needs and that environment have become more complex, humans have developed more and more complex systems of organisation and navigation to be able to cope. Given that our psycho-physical make-up is not capable of change at an equally rapid rate, inevitably there will come a time when the complexity of our magnificent constructions will become incomprehensible to us. They'll cease to help us and begin to frustrate us. Some people would say this point has already been reached. Certainly, there's a general realisation that if we're to continue to benefit from our own ingenuity, we must take steps to eliminate all unnecessary complexity. We must save our limited capacities for dealing with things that matter.

The concept of Harmoniopolis I've outlined here aims to provide people with a quasi-urban experience in which the 'natural' or intuitive elements of human life are maximised while still maintaining the benefits of a complex society. The elements of our urban Ur-environment are incorporated (light, water, flora and fauna, access to other communities, etc.) and the internal organisation is made comprehensible through natural, iconic modes of signage. The result is a harmonious whole in which sustainable growth (i.e., improvement) is possible, since it takes account not only of the needs of the homo sapiens of 2000 AD, but also of those of our forebears from 2000 BC – people whose deepest desires and thought patterns we still carry with us as our own.

# Uniformity, Diversity and the New Nomads

Will archaeologists looking back at the late twentieth century call it the Age of the Walkman? Certainly, the street scene in many parts of the world would be incomplete today without it, and many will surely be around to be dug up centuries hence. Indeed, we're increasingly surrounding ourselves with such miniature, personalised electronic products. The pace of this development has recently begun to accelerate rapidly, thanks to new technologies such as the integrated circuit or chip. The handheld computer, the pocket-sized television, mini stereo-systems, and so on, can be seen as the latest steps along this path, a path which is leading to the re-interiorisation of the extended powers that resulted from earlier exteriorisation. The personal stereo, for example, almost allows us to incorporate our new expanded and exteriorised faculty of hearing into the body – right alongside our natural hearing. Personal organisers are gradually allowing us to incorporate the expanded power of memory afforded us by the computer into the body, or at least into the breast-pocket. Mobile phones and many other portable products are moving in the same direction, giving us more power and more independence. How are these developments affecting our lives, how will they affect them in the future?

**A multicultural world**   It's noticeable that as people become more familiar with miniaturised portable tools, they also become more excited about using them to explore the world. These tools open up new horizons. Just as railways, automobiles and aircraft have made it possible for us to venture into new territory, so today's miniature tools, with their expanded software, are enabling people to explore new possibilities in their lives. And as people realise just how much they can do with miniaturised products, the more they will actually *want* to do.

The same development can be seen in the cultural field. Globalisation and greater mobility are making people more aware of other cultures. Formerly tightly-defined ethnic and social groups are becoming looser and less inward-looking. As consumers, too, people are now cherry-picking from a whole array of behaviours, allegiances or cultural artefacts, mixing them together into a unique personal style. The more our societies become multicultural and the more we see of what there is in the world, the stronger becomes our desire to experience it all. We want to go everywhere, do everything, feel everything – even be everyone. Life is arguably more of an adventure now than it ever was when new continents still remained to be conquered.

**Continuous interplay**   Won't this development lead to the creation of an undifferentiated global culture and, ironically, to the demise of the very diversity that inspired the spirit of exploration in the first place? I don't believe so. There seems to be a continuous interplay in human existence between uniformity and diversity, between communal and individual behaviour, between conformity and freedom. We see this interaction in the historical development of many of the objects that surround us in our daily lives. There seem to be three distinct phases in this evolution.

Velo Mono: a concept for a handheld computer with
touch-screen and telecommunication facilities

**From technical to informational globality** The first or oldest phase is what we may call the era of technical globality. Archaeological evidence shows that implements of similar size and function developed separately in different parts of the world. This is presumably because their forms were determined largely by the universal constraints of human physiology and psychology and the laws of physics. In other words, the homogeneity of these earliest human artefacts was the result of our common heritage as homo sapiens in the physical world.

In the subsequent course of human history, however, different ethnic groups took these basic shapes and embellished and reworked them to suit their own socio-cultural needs and tastes. This second phase may be called the era of ethnic blocs. Take, for example, the development of an eating implement. Despite a basic similarity (a stick held in the hand), the Chinese chopstick is quite different from the western fork. And then again, the English fork of, say, the Victorian period is quite different (though less fundamentally so) from the Italian or German fork of the nineteenth century. The same principle still applies today: an American refrigerator, for instance, is different from a German or Italian one, each responding to the needs and tastes of the ethnic group being catered to.

Ethnic diversity of this type was originally – and to a considerable extent still is – geographically limited to what we may call the tribal homeland. However, certain ethnic cultures have now been spread over a wide area through colonisation, neo-colonisation and trade. This has led to the rise of a sort of international culture, comprising particularly the more recent technological products, such as TV sets, computers, microwave ovens, and personal stereos. Such products have come into existence since the emergence of global markets, transcending regional boundaries or ethnic blocs. We are now, then, in the third phase, which we might term the era of informational globality.

Although we see evidence of this globalisation of culture all around us, ethnic diversity is still alive and well, and with a great future ahead of it. We shall increasingly use it to express not only our ethnic origins but also our own unique identity and through it achieve a measure of self-fulfilment.

**Re-interiorisation** I just mentioned examples of how a basic functional object may be embellished by various cultures in different ways and at different times, while still retaining its basic function. However, I also referred at the outset of these reflections to the partial re-interiorisation of certain functions, citing the examples of the personal stereo, the cellular phone and the handheld computer. Continuing miniaturisation will mean that such functions will, to all intents and purposes, 'disappear' from view: they will be functions without form. What will then happen to today's tangible carriers of these functions? Will they simply vanish along with the functions?

Perhaps: a trawl through history turns up many such carriers, lost because they had been bleached of their function. Crossed garters for medieval men vanished when improved tailoring led to the creation of tights and made them unnecessary, for instance. Alternatively, the carriers may become fashion accessories (as cellular phones have already become for some). Passing through a combined functional-decorative phase, they may ultimately survive as purely decorative objects. This is what became of the male neck-tie: starting out as a means of fastening the collar, it gradually became more decorative, not reaching its present purely totemic stage until the nineteenth century. Women's brooches, now almost completely decorative, similarly began as a means of securing clothes around the body, passing through a combined functional-decorative function before becoming the pure adornment they are today.

Of course, for the foreseeable future, miniaturised functions will not be entirely intangible. They will be incorporated into our clothing in such a way that we shall not notice their physical presence except when they are in use. But

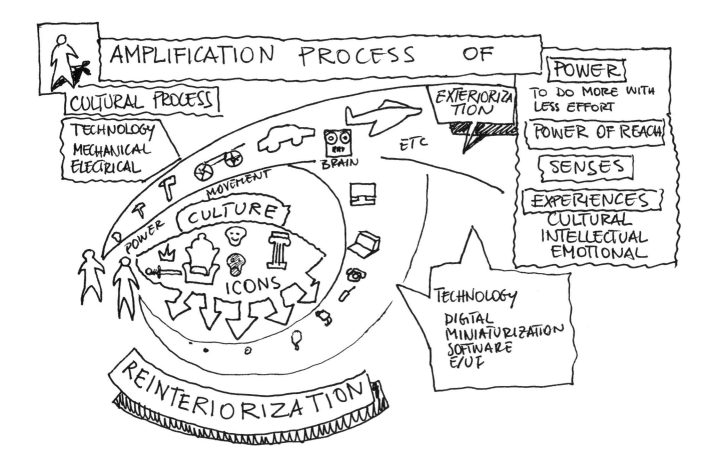

will this incorporation leave that clothing unaffected? Again, perhaps. We often conceal new functions – lighting fixtures are often 'built in', for instance. But the new 'invisible' functions may equally well be given visible form, not because it's a necessary consequence of their physical presence, but simply in order to announce that presence. Such indexical forms, originally pointing to the hidden function, may eventually even redefine the aesthetic of the original object. Sports shoes, for instance, are an adaptation of conventional footwear, and have developed an aesthetic of their own. This aesthetic is now spreading to affect other casual and even formal footwear.

Another reason for giving a new function a visible form even when it's not necessary is in order to 'explain' the new function to users. This is done by drawing an analogy with an older function linked traditionally with a particular form. Thus, the weapon wielded by Luke Skywalker in *Star Wars* had a technology unknown to us. Yet it was given the physical appearance of a sword or wand (albeit an illuminated one) whereby we were able to understand its function. Similarly, when electric kettles first came onto the market, they resembled the local form of the object normally placed on the fire to heat water for drinking – in some cultures that was the tea kettle, in others the coffee pot. This blending of the old and the new makes it easier for people to understand and accept the new.

**Portable and personal**  Such developments have implications for the way we as designers have to approach the problem of designing products. First of all, the need for diversity within globality means we need to concentrate on producing mass quantities of products which are, paradoxically, essentially the same and yet essentially different. To be economically viable, products need to share a common technological core. At the same time,

THE UNCONNECTED HOUSE
"LA CASA ISOLATA"

THE CONNECTED HOUSE
"LA CASA CONNESSA"

to be truly satisfying, they need to be capable of a high degree of personalisation. Individuals need to be able to use the product in precisely the way they want; and this requires very versatile software. Secondly, the external appearance of the product (its hardware) also has to be highly adaptable so that it can reflect people's personal choices, conveying exactly the cultural qualities they wish to embrace as their own. And it's precisely products which represent re-interiorised functions, products that nestle close to us and can be carried with us wherever we go that are ideally suited to becoming highly personalised in this way. Thirdly, a product's personal quality must also extend to the functions it performs. These must be functions that people feel are important and relevant to them, that help them realise their goals and aspirations. The products we develop should therefore not be ends in themselves but rather creators and bearers of knowledge, services and emotions; they should not merely convey image but should allow people to express their true identity.

And finally, products should allow their users to perform those relevant functions in an entirely natural and unhindered way. This not only calls for intuitive interfaces and control modes, using voice and touch, but also minimal presence. It means requiring a minimum of effort on the part of the user for a maximum result. We'd like, for example, to communicate without either walking to a phone or fumbling in a pocket for one. We'd like to be able to experience sensory contact without having to travel. We'd like, in effect, to be ourselves, able to do everything, everywhere, both at home and on the move. Surely, in a society in which wearing clothes is the norm, the unobtrusive incorporation of relevant functions into clothing or fashion accessories would be an ideal way of taking a quantum leap towards full empowerment and unlimited reach.

**The connected society**   Before we get carried away, however, we shouldn't forget that the steady increase in power and reach that tools have already brought humankind is paralleled by another development – the increasing interdependency of human society. Since the very earliest times, humans have gathered together in communities because such alliances allow us to do more than we could do on our own. Today, everyone is dependent on someone or something else. Travellers, for instance, depend on the airline company to take them from A to B; the airline is dependent on its staff and fuel suppliers; the pilots depend on air traffic controllers, who in turn depend on their instruments, and so on.

We're quickly becoming integrated into a vast extended community, or rather, set of interconnected communities. Virtual connections will link people into all sorts of groups besides those in which they find themselves physically. Part of our responsibility as designers and makers of electronic products is to see that these new contacts are effective and satisfying. No one must be excluded from this interconnected world. If all are to benefit from the opportunities technology offers them for expanding their individual intellectual and cultural possibilities, their power to achieve things, and their ability to reach out to other places and people, we must make it easier for them to use the inevitably sophisticated electronic equipment which will increasingly characterise our information-oriented society.

"GENTE SOLA.,
"LONELY PEOPLE.

"UOMO CONNESSO.,
"CONNECTED MAN.,

**Moving and living**  The linking up of the world into ever-finer networks is an accelerating process. The more we are joined together, the more we realise what we could do and experience. And the more we realise what is possible, the more we shall want to try it. Already, many global businesses operate around the clock. The Internet is allowing people to make friends on the other side of the globe – people who, sooner or later, they might want to visit. As we rush around experiencing more and more, time will be at a premium and we'll want to make optimum use of it, including while we are on the move. Communication, vital in an interconnected society, will be an important part of mobile activity. Miniaturisation, cheaper satellite links and better digital compression are already making it possible for people to communicate with each other in almost any medium wherever they are.

This highly convenient and direct mode of communication will have unexpected knock-on effects in many areas of life. Children, for instance, will be freer to play where they like, since improved monitoring technologies will allow parents to check on their whereabouts from a distance. And as better communication compensates even more for the time delay that physical travel inevitably involves, a wide variety of new benefits will emerge, ranging from medical diagnosis and treatment at a distance, to the rapid location by rescue services of people in danger.

**The new nomads**  The re-interiorisation of our expanded natural powers, the rapid spread of the multicultural society and the increasing interdependency of everyone on the planet are all making possible a new-old mode of life – that of the nomad.

Thousands of years ago, many human societies chose to abandon their nomadic lifestyle and settle down as farmers. We have inherited their legacy. The itinerant and the static way of life were always at odds: the restless searching and rapid upheavals that characterise the life of the nomad are difficult to combine with the slow business of putting down roots and patiently waiting for the cycle of seasons to yield its fruits. For our ancestors, it was, of necessity, either one thing or the other. But humans are complex creatures. In all of us there is something of both the nomad and the settler.

Now, at last, we have the promise of being able to combine them. We won't need to abandon our static life when we travel, nor forego the joys of adventure and movement when at home. Virtual travel will bring the world to our door, while portable products will enable us to maintain and enjoy our settled existence as we explore the physical world beyond. We'll take with us all the comforts, joys and securities we've come to value since farmers first sowed their crops, combining them with the excitement of travel.

# Chocolate for Breakfast

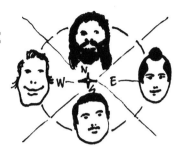

Chocolate for breakfast is not everybody's taste, and some people – like me – even find bacon and eggs hard to take first thing in the morning. But a couple of centuries ago for rich families in Europe it was very much the muesli of the day. Thinking about the task facing designers in the twenty-first century, I'm reminded of a poem by the Italian writer Guiseppe Parini called *Il Giorno* – 'the day'. It describes a day in the life of a well-to-do family, living a very comfortable life in a sheltered world, quite unconcerned about what conditions were like for those less fortunate than themselves. And, of course, they had hot chocolate for breakfast. Theirs was a world of comfort and privilege. It was part of a Western European idyll, one that was intensified by the Industrial Revolution, only to be finally shattered in 1914 when the First World War broke out.

It was the world par excellence of laissez-faire capitalism, pursuit of profit, ruthless industrial competition and imperial adventures – and it led directly to the Battle of the Somme and the futile horrors of trench-warfare. The subsequent imposition of impossibly heavy war reparations on Germany immediately sowed the seeds for the Second World War, while the Russian Revolution – a different reaction to unbridled privilege – contained the germ of the Cold War and balance of terror which were to follow three decades later. Now, the Iron Curtain has rusted away, and here we are, in 1993, witnessing in eastern Europe the latest – but certainly not the last – link in a long chain reaction.

In the relatively comfortable setting of the Industrial Triad countries – North America, Western Europe and Japan and Australasia – we're metaphorically sipping our chocolate for breakfast, while the masses outside begin – with increasing insistence – to knock on the door, demanding their fair share of the goodies we've been privileged to enjoy for so long. In other words, as we approach the millennium, we're being confronted with the effects of a gross imbalance in material circumstances around the world. Our idyll, our sheltered existence, is beginning to crumble. And not only because we're being challenged by our fellow human beings, but also, of course, because our physical, natural environment is showing severe signs of strain.

**Restoring the balance** If we're to look forward to a period in the future in which we have a stable environment and can pursue sustainable growth, then we must try to restore the balance in both our natural environment and in our social and cultural environment. We must try to create a world of 'happy objects and happy people', a 'Paradise Regained' in which people live at peace with each other and their environment. But this ideal world is not a static worl: it is one in which individuals continue to grow and achieve personal fulfilment. It is, in fact, a world of sustainable growth.

But how to get there, from where we are today? At the moment, we're still living with the legacy of the Industrial Revolution and the 'get-rich-quick' mentality which brought the world of our chocolate-loving Italian family to such a bloody end and continued to reach out and grab us with its tentacles for decades beyond. Manufacturers still compete on the basis of quantity, with added extras which may temporarily boost sales but lead ultimately to frustration. The alternative is to work towards quality. The quality I mean is difficult to describe and yet so easy to recognise in one's own life when it occurs: having time to spend with loved ones, or explore interests or help others, for instance, rather than collecting money and status symbols. A sustainable society requires us to adopt this sort of mature, quality-based outlook.

**Society, government and design**  As designers, we're in a position to help our societies move in that direction. But we need to take proper advantage of that privileged position. We – the design community as a whole – need to be actively involved in making our ideas and thoughts known on the macro-level, the level of governmental institutions, international organisations and society at large. At the moment, nothing much is happening on the macro-level in relation to design. The role of government in industrial design is minimal. Neither ministries of culture nor ministries of industry seem to see design as being within their domain. It falls between two stools.

There are occasional, notable exceptions, of course. We shouldn't forget, for example, the British government's initiatives in the early nineteen-eighties to awaken industrial management and the public at large to the importance of industrial design. It's worth noting, though, that they were clearly not motivated primarily by cultural considerations but by economics – design as a way of boosting sales.

Others have also begun to wake up. In 1993, the Japanese Ministry of International Trade and Industry announced plans to promote new design policies throughout the Pacific Rim region. And around the same time, the Clinton administration in the United States contacted the design world there to discuss relevant issues with them. But despite these occasional stirrings, governmental interest in design still seems to be the exception rather than the rule. In fact, most governments don't seem to see any connection at all between industrial policy – a matter of major concern for all countries – and the cultural development of the people they're governing.

It's not enough for governments to just sponsor conferences and competitions or set up committees. Everyone – every consumer – needs to be made aware of the vital contribution good design practice has to make towards the realisation of sustainable growth. Of course, there's actually no reason why governments should not get involved. Many of them, for example, see it as one of their tasks to regulate television broadcasting, to safeguard the quality of programmes. In other words, they're involved in regulating software. What then of all the other software that's around? Besides video tapes, there are an increasing number of consumer products available which have a software element, and some are, in fact, predominantly software. Why aren't governments giving thought to these? And this is only one area we could mention.

**Design is a political act**  It's surely the duty of the design community to make sure that both governments and society at large take an interest in design issues. And if they're to take an interest, we have to make sure that they're well-informed. To do that, we need to become conscious of our political clout and significance. Design is a political act. Every time we design a product we are making a statement about the direction the world will move in. We therefore have to continually ask ourselves: Is the product we're designing relevant? Is it environmentally responsible?

The solutions we choose are political decisions, not design decisions. And political decisions are steps out into the global arena. Designers must become aware of their power. They should not underestimate their power to lobby and persuade and cajole; they should explain to the world how design factors play a vital part in affecting the quality of life; they can educate and advise, coaxing people away from the worship of quantity towards the pursuit of quality in its broad sense. Certain national groups and individuals are already doing important work in this area, but it is unity that creates strength, and, if the voice of design is to be heard above the clamour produced by proponents of the short-term economic quick-fix, it's strength on a global level that we need.

**Accelerating change** Once governments, organisations and the general public start to understand the issues, this will inevitably trigger a snow-ball effect. Change will accelerate. Governments can establish industrial guidelines for maximising the relevance and cultural value of products, and they can develop ways of raising public awareness of such issues. Then, when we talk to colleagues from other disciplines, we'll be backed up by friends in high places, including that friend we'd all like to have backing us up – public opinion. The global design community should be involved with governments as partners, talking to ministers of culture and industry, proactively generating a new awareness of the wider importance of design issues, showing that a vital link can be made between industry and culture, that the two *can* mesh together in a mode of competition based on quality rather than quantity, and that this is the blueprint for realistic sustainable development.

Time, however, is of the essence. We're involved in a race against the environmental clock. Look how unwilling some governments were to commit themselves to rapidly phasing out the use of CFCs. Even though we can all see, in the now familiar computer pictures of a multi-coloured whirling world, how those ominously dark holes in the ozone layer continue to grow and grow. Changing people's minds and counterbalancing the attractions of short-term thinking are things that take time. After all, the spirit of laissez-faire entrepreneurship which inspired the Industrial Revolution in the nineteenth century was alive and well in the West as recently as the nineteen-eighties under the names Thatcherism and Reaganomics. In other words, we shouldn't underestimate the weight – the deadweight – of the status quo.

What we're fighting against is the present-day equivalent of the myths, superstitions and power structures of the Middle Ages. In fact, we're on the eve of what could be called a New Industrialised Renaissance – a revaluation within our high-tech society of the human experience in its broadest sense, with genuinely relevant products meeting our true needs and aspirations, expanding our awareness and reach, deepening our understanding and power, expressing our identity, all within a healthy and sustainable natural, corporeal and social environment. The propagators of new humanist and liberating ideas... half a millennium ago – people like Galileo, Luther and Erasmus – were not given an easy ride by the establishment. But if they hadn't taken a stand for what they believed in, much of what we've achieved over the last five hundred years, imperfect as it may be, would not have been possible.

Designing for the next millennium, designing for the New Industrialised Renaissance, demands that we in our turn take a stand right now for what we believe in. Because if we delay, if we're content to hesitate on the sidelines, we'll be too late. We'll just watch as our idyll disintegrates around us.

**Guidelines for the Design Community** A number of years ago, the British Design Council billed designers as 'the New Alchemists' – "adding the business ingredient that sends sales soaring." But designers are not alchemists. Alchemy belongs to the Middle Ages and to the get-rich-quick mentality. The riches we're after are less tangible than gold but of greater value. And they won't come from some magic formula. They'll have to be worked for, systematically. To steer us in this effort to attain our Paradise Regained, the design community needs, I suggest, a set of guidelines drawn up on an international basis for industrial growth, product generation and corporate business policies.

**The environment** Obviously, environmental concerns should feature strongly, particularly in relation to the use of materials and recycling. Where possible, quality should be emphasised above quantity, so that we don't waste finite resources on unnecessary duplication or superfluity. Furthermore, a sustainable balance should be established between quantity and quality in all parts of the world.

**Economic imbalance** The economic imbalances between North and South globally, between East and West in Europe, and, indeed, between the rich and poor in many of our own countries should not

THE NEW WORK SHOP.

19/1A 93

WHEN MICHELE AND MYSELF MET IN MILAN I HAD ALREADY THE FEELING HE WOULD HAVE BEEN VERY ENTHOUSIAST OF MY PROPOSAL - SINCE SOME MONTHS WAS CROSSING MY MIND THE IDEA OF THE NEED TO ESPRESS NEW OPINIONS ON THE MODEL OF WORK AND ON THE TOOLS AVAILABLE TO DO IT. THE IMAGES OF MY GRANDFATHER LAB AND OF THE SHOEMAKER PELLEGRINI ALWAYS GAVE ME A FEELING OF "LABORIOSA SERENITA" - A QUALITY WE SHOULD AIM TO AS THE NEW QUALITY FOR SUSTAINABLE DEVELOPMENT. MEN WORK IS TO SUPPORT MEN CULTURAL GROW AND WELLFARE. IT SHOULD NOT BE THE CONTRARY AS IN MANY CASES OF TODAY REALITY WHERE MEN EXIST TO ALLOW INDUSTRY TO EXIST - HOW TO FIND THE BALANCE. HOW TO SUGGEST MODEL OF "LABORIOSA SERENITA" SUPPORTED BY ADEGUATE AND RELEVANT Tools. IN VIA GOITO MICHELE AND MYSELF AGREED ABOUT THE NEED TO TAKE A PAUSE IN OUR DAILY "LABORIOSA SERENITA" TO THINK AND

only be recognised, but we must also do something about them. However, as long as the pure-profit Quantity Doctrine dominates, few companies will voluntarily undertake seemingly risky investments in poor countries. Yet such investment is absolutely necessary. It is generally understood now that assistance should not all be in the form of hand-outs or loans from governments. This just results in more dependence. But if we leave everything to the laws of free-market economics, we shall have to wait forever before those countries are in a position to sustain their populations adequately on the basis of their economic activity.

Why not then some sort of positive discrimination? Companies could be encouraged to develop investment policies aimed at enhancing global sustainable development. Making products specifically geared to needs and circumstances of developing countries, for example. Developing countries will be important markets for more sophisticated products later, but only if they've been able to develop in an economically and socially balanced way. If the world as a

MEDITATE ON THE ISSUE. THIS, WE
DECIDED, WILL BE A FORM OF
POSING QUESTIONS, BURNING QUESTIONS,
TO OURSELF AND TO THE DESIGN
COMMUNITY ALL — THIS, WE DECIDED,
WILL BE A FORM TO SEARCH FOR
ANSWERS — THIS, WE DECIDED,
WILL BE A FORM OF A DIALOGUE
WITH SOCIETY — THIS, WE DECIDED,
WILL BE AN ACT OF DESIGN —
DESIGN INTENDED AS A POLITICAL
ACT — DESIGN CONSCIOUS OF ITS
IMPACT ON THE FUTURE FORMS
OF WORK LIFE QUALITIES —

WE DECIDED TO HAVE A WORKSHOP
IN HOLLAND WITH SOME FRIENDS
AND COLLABORATORS — WE NAMED IT
"THE NEW DESK LANDSCAPE "
WE DREAMED ABOUT IT — TODAY IS
REALITY — IT WAS DESIGN AS A POLITICAL
ACT — IT WAS DESIGN AS A ACT OF LOVE —

whole is to form a stable community of mutually sustaining trading nations, then it's in the interest of all of us that such balanced development is given a chance to take place.

Essential but unprofitable products and services    A related problem on which we should develop a standpoint is that of who is to make things or provide services which are not directly profitable but which are nonetheless socially and economically necessary or desirable. This might include public utilities and public transport, but also the research phase of new technologies in universities or industry. State support or control of such activities has been tried in a number of countries this century, but in most of them this approach has been found wanting and they're being rapidly privatised again. Is this the right way forward – or is it in fact a swing of the pendulum back to the past? Is there perhaps some other solution? Shouldn't we perhaps be moving towards a more global approach? Certainly, global agreements in a number of these areas could render investment less risky and therefore more attractive. Alternatively, some sort of globally agreed levy imposed fairly on companies worldwide could be used to support such activities wherever they were needed in the world, as part of a programme of what we might call World Profit Management.

The benefits of such a programme wouldn't only be material. Designers would also have a professional interest in the maintenance of certain unprofitable activities. Often enough, such activities ensure the continuation of poor or isolated communities, together with their unique cultures – communities which might otherwise die, as younger inhabitants leave in search of comforts more readily available elsewhere. And as each community and culture dies, human diversity is diminished. Look at how agricultural communities all over the world have abandoned the land for the bright lights and comforts – the electricity, the water and plumbing – of the cities, only to end up living in squalor in shanty-towns. The current flood of refugees into Europe and the United States from the former eastern bloc, Africa and Southeast Asia is simply an updated version of the same story. Preserving human dignity and diversity is every bit as much a conservation issue as preserving pandas and rain-forests.

Unemployment   Many of the products we produce are responsible for putting people out of work. The march of technology inevitably brings pain with it, the pain of transition. Weavers were put out of work in eighteenth-century Britain when the automated loom, the Spinning Jenny, was introduced, and later, when the steam train arrived, many a coachman and ostler must have ended up on the street. Most western countries now have some way of alleviating the misery of being without work through unemployment benefits. But during the past few years we've been able to see that this only solves a small part of the problem, providing the basic material needs. It doesn't address the frustration and damaged self-confidence which losing your job entails. What we might suggest as a way of tackling this problem which maintains the quality of life, is this. When developing a new product, companies should design not only the product itself but also a scenario of how the effects the product will have on people can be accommodated – and not only accommodated but positively exploited.

This means that companies will have to think ahead in socio-cultural terms, to beyond the moment of sale. What happens after the product has been sold should not be left to chance but be designed. For example, if we're introducing a new automatic word-processing system, we should provide a scenario to take account of the fact that the machine will probably be replacing a secretary. The scenario might include the provision of training for the secretary so that that person can perform some new function for the company – an 'added value'. In other words, the manufacturer would not only be providing the hardware, but would also be providing an element of socio-cultural engineering. He would in fact be undertaking the conservation of human resources. The resultant situation would be one in which there were no victims, no-one being told that, as the latest euphemism has it, they're being "given the chance to seek a new career opportunity elsewhere." There would only be winners. In fact, governments might even be encouraged to make the provision of such a human-resources-conservation scenario obligatory. It would save the community the burden of paying unemployment benefits and it would maintain the dignity of the employee.

The rise of services   Designing such scenarios for new products is only part of a much larger trend away from the hardware aspects of products towards their function as the carrier of services. With the increased role of software in products and the need to conserve the material from which the hardware is built, this trend seems likely to continue. Upgrading software is more environmentally-friendly than bringing out a new model, even if certain hardware elements of the product have been recycled. And the next logical step in conserving materials would be for us not to own the hardware, but only to hire it when we need it. In other words, the centrality of the hardware, the traditional object-product, in the mind of the consumer will diminish, and the service will become paramount. The concept of property may even change. And certainly designers will find themselves being more designers of services than designers of objects. Although the full ramifications of this trend have yet to become clear, it's surely a development which a politicised design community should not only be aware of but have thought through and be ready to steer.

**No longer invulnerable**  The days when you could produce a pure techno-fix product and put it on the market without considering its potential implications for the world at large are over. Chernobyl saw to that. Our world is so fragile that we can't afford any more mistakes – we must get it more or less right the first time. This applies equally, if not more so, to the effect of products on people. We've worked out how to measure objective quality in products. Now we have to face the problem of measuring socio-cultural quality. We certainly used to have the impression that we were invulnerable. There was no limit to humanity's technological progress. But the hole in the ozone layer has shown us that our environment is under threat, and that we, too, are at risk. This realisation must surely lead to a new synergy, linking environmental conservation with the conservation of human socio-cultural values. The chickens have finally come home to roost – or even roast. We have lived a life of wild abandon, but we're now sobered as we notice the first symptoms of debility creeping up on us.

**The role of the media**  We also need to enlist the media to our cause. This is crucial, because if the general public fail to understand what we're doing, if they fail to appreciate the larger significance of our new type of product and continue to make their choices on the same basis as in the past, then we shall not succeed. The public need to be educated and the media can do that – but they'll need input and encouragement from us. We need to make them aware of the new multidisciplinary approach to design – High Design – and why it's so important. We need to make them aware of their educational function, their vital role in accelerating the process of creating a sustainable future for us all.

**Yes, but...**  Some designers may say to all this: We're not politicians; our business is designing. Certainly, our primary responsibility is to influence the field we work in and the processes we're involved in. We have to develop strategies in that context, to make sure that we're carrying out our design-work in an ethical way. After all, that's ultimately in our own interest, because if things continue along the present quantity-oriented path, in a saturated market, there'll be fewer jobs for designers. But we must also be in position to see beyond our own garden wall, and see things in perspective. To talk about design issues on different levels and in different contexts. The fact that this may sometimes be a complex and difficult task should not frighten us off. On the contrary, it should inspire us to find a way of organising ourselves so that we can address these issues effectively both at the micro and the macro levels.

If we recognise that we're not merely technical operators but cultural operators as well, manipulating, expressing, playing with and delighting in human culture in all its diversity, then we must grasp the responsibility and make our voice heard in the courts of the decision-makers. Maybe we'll succeed in convincing them, or maybe we won't. But we have to try. What else can we do? Sit back and wait for someone else to say it all for us? If we do, we'll wait a long time. Too long. The future is created by those who accept – no, take – the responsibility for it today. What we certainly must *not* do is to just sit back – and continue sipping away at our breakfast chocolate...

# In Search of a Sustainable Society

Since the fall of the Berlin Wall in 1989, our perception of the world has changed radically. Before that, for most of us the 'known world' was limited to 'our' side of the Iron Curtain. True, we were aware that life existed on the other side, but we knew little about it. Only the external facade of the monolith impinged at all upon our daily lives. Now, we've come to see the world as a single system. Poverty on the streets of Moscow or Mexico City, the problems of Bulgarian or French farmers, the successes and failures of Czech or Chinese industry are all now perceived as things which can affect us, either in the short or the long term.

**Out of balance**   Although the world may now constitute a single system, it is a system that is clearly out of balance. Every night, through the medium of television, we become eye-witnesses to what is going wrong around the world. Fighting and suffering, economic problems, political crises and social conflicts – the details may change, but whether in Bosnia, Rwanda or Chechnya, the effect is the same. This daily confrontation with the suffering of others, with the fragility of existence itself, makes us uneasy. On top of this, we're also made aware that the physical environment of our planet, and therefore ultimately our own survival, is under serious threat. Oil spills along the coast of Alaska, the Chernobyl disaster, the depletion of the ozone layer, endangered wild life, the destruction of the rain forests, and so on, and so on.

**Sombre environment**   Our knowledge of events around the world creates a sombre mental environment from which we can never escape for long. The younger generation today is growing up with this, and it is surely one of the factors that differentiates them from their elders. While this is most obviously the case in the industrialised world, it's becoming increasingly true of youngsters in the third world, too.

Perhaps if such instability on a global level were balanced by stability on a local level, it would be easier to deal with mentally. But life at home and at work are also changing at an increasingly rapid rate. The individualisation of society, which began several centuries ago, has loosened national and family ties. Increased mobility has weakened our sense of belonging to a particular place. And work, if there is any, calls for flexibility: and certainly few people can expect to stay in the same line of work all their lives.

**Transversal**   Maintaining a sense of identity and a sense of stability through all this is not easy. Yet this is what human beings must do if they are to retain their equilibrium, self-respect and, above all, their humanity. People react to this problem in different, often totally contrasting ways. Some adopt a position at the extremes. At the one end of the spectrum, there are the activists, those who devote themselves to righting the wrongs they see, arousing awareness of them in us, too. At the other end, we see the escapists, ignoring the larger problems, turning inwards and becoming preoccupied with superficialities. But most of us find ourselves somewhere in the middle, exhibiting a transversal attitude. At times, we're concerned and affected by what we read and see. At other times, we turn aside from serious matters, relaxing and indulging in luxury as we ignore for the moment the problems of the world.

**Fake**   But it's only a temporary deflection: we can't turn our backs on them for long. In fact, as I see it, people are increasingly rejecting things that seem empty, without meaning. It's as if they're deciding that it's time to concentrate on the important things in life, on basic values and qualities. There's no room for frills, superfluous ornament, or pure show. In the face of the serious problems faced by the world, it just seems fake, an obscene waste of resources, time and psychological energy.

**Sustainability**   There's a growing awareness that the society of the future will need to be a sustainable society. We've grown up in a world based on the Modernist vision of relentless industrial progress. Though it looked ahead to future benefits, it did so without casting a glance to left or right – and certainly not behind – to see what havoc it was creating as it bulldozed on. The society that has been created is, we now see, far from sustainable: continuation along the same path will lead to the ultimate catastrophe. Instead, we're beginning to appreciate that we need to take a more holistic, more global view of progress, as we work towards a sustainable society.

A sustainable society would be one capable of developing new, renewable sources of energy, of minimising the use of non-renewable resources, and respecting the natural and man-made environment. It would also be a quintessentially human society, in which everyone will be able climb the ladder of needs already indicated in ancient times by Plato and more recently by the psychologist Abraham Maslow. Satisfying first physical needs, then intellectual and finally spiritual needs, they attain the ultimate form of self-fulfilment – what Maslow called 'self-actualisation'.

This would be a world where 'travelling' is as important as 'arriving'; where the process of making a cup of coffee is as important (and enjoyable) as the cup of coffee itself. It would be a place where our psychological need to expand our experience and to achieve goals would be fed and met. It would also recognise that, as social creatures, we place a high value on qualities such as peace, peace of mind, love and fraternity, and it would accordingly encourage these qualities in everyday life.

**Are we looking at the answer?**   Exactly how such a utopian society is to be attained – starting from where we find ourselves today – is, of course, unclear. Simplistic doomsday scenarios which call for massive 'backtracking' and the surrender of many of the conveniences industrialised society has come to depend on would probably also require the surrender of many of the individual freedoms. On the other hand, gradual evolution will almost certainly only bring change too late.

Perhaps a viable answer is actually staring us in the face. What are the most striking phenomena in the world today besides those sombre aspects I've mentioned? I'd say that one of them at least is surely the rise of information technology. Information technology was a natural development within industrial society. It basically continues the process of automation which began many centuries ago, though it does so in a way that takes it along a radically new path and at a vastly accelerated pace. It has been growing, like a natural system, in a scattered, unstructured fashion, without any overall motivation to give it direction.

But if we now place the needs of a sustainable society alongside the opportunities offered by information technology, we see a match. It's a match that's so close that it calls out to be exploited. If we can gear the future direction of information technology towards the goal of realising a sustainable society, we'll provide each of these two endeavours with what it needs. Information technology will acquire a clear-cut and worthy aim, and sustainability will have a chance of becoming reality. What could be more significant for all our futures?

**Information technology** Let's consider the potential points of contact. In the first place, imaginatively exploited, information technology can answer many of the environmental requirements of a sustainable society through its highly efficient use of energy and materials. It can therefore reduce our impact on the environment considerably. Thus, miniaturisation, made possible by chip technology, allows us to cut down on the use of material resources. Software, by being upgradeable without the need for extensive production plants, enable substantial energy savings to be made. And when linked to telecommunications technology, software also reduces the need for transportation and storage space, with further savings of non-renewable resources. In the second place, information technology can offer solutions to some of the social, cultural and intellectual needs of people in a sustainable society. Computer networks, such as the Internet, belong to a world without boundaries, a world in which news travels faster than gossip in a village, with new 'virtual' communities, in which everyone can chat over the garden fence and have their say. Finally, by making more information available, this technology will make finding satisfactory solutions to problems, small or large, considerably easier.

**The New Modernity** In this light, the interaction between the needs of a sustainable society and the possibilities offered to us by information technology can be seen as something of a 'holy alliance', a coincidence of historical chronology which can give rise to a new, more promising world. In contrast to the historical period known as Modernity (i.e., the Industrial Age now drawing to a close), information technology, given direction by the urgent need for us to attain sustainability, can lead us towards a 'New Modernity' – an age in which we can continue to believe in the idea of progress, but where that idea has a radically different content, one based on quality instead of quantity.

**Who creates the future?** The future is not something that is created by the efforts of only a few individuals, companies or governments. It is the result of many contributions, of momentums initiated in various places, of decisions whose impact may not be immediately apparent, but which ultimately and jointly can become highly significant. Therefore, without running the risk of being charged with hubris, we, as producers of goods and services, and as designers, can – indeed, must – make our contribution in whatever way we can. Creating a new household appliance or multimedia application may seem a minor venture in the grand scheme of the future, but that doesn't mean it will be without effect. Everyone must take positive action with a clear goal in mind in those areas in which they have expertise and influence; joining with others who share their vision of a sustainable society to try to guide things in the right direction. This is, in essence, the ethic of the new era.

**What is our duty?** Specifically, the more we become aware (and can make others aware) of the potential effect that information technology has for improving the quality of life, in terms of advancing the arrival of a New Modernity, the more we'll be able to direct our work and accept our responsibility to further this goal. On a very practical level, such a goal provides a motive that can help us determine what services we might offer and what products might most clearly affect people's lives positively. To make products and services that move us all towards a sustainable society, we not only need to make them environmentally friendly, we also need to ensure that they focus directly on basic human values and qualities, advancing self-actualisation and the higher cultural and social virtues that have always been recognised as bringing society the greatest good.

# The Metamorphosis
# of Products

All too often, people see the comforts of our technological society and the needs of the environment as being antago-nistic to each other. They see ecological design as implying that we must give up many of the handy inventions of the past century and take a step backwards. But design can affirm and show that such a pessimistic scenario is not inevitable. It should be possible to create ecologically-designed products which are at the same time attractive and charismatic enough for people to buy and to be perceived as a contribution to improving the quality of life. This is an ambitious goal, certainly; but it's one worth pursuing, both on ethical and economic grounds. In accepting this challenge, various paths can be followed, strategies which could be pursued to realise products compatible with sustainability. Let's con-sider here four strategies in particular. We'll call them *New Heritage*, *Rent Line*, *Improve Now!* and *Relevant Functionality*.

**New Heritage** One major factor burdening the environment is the relatively short life span of many products. We need to make products which last longer, and – more importantly – which people also want to keep longer. To understand how to do that, we need to consider what it is about modern products that makes it easy for their owners to throw them away; and to recall what it was about objects in the past that made people keep them and pass them on to their children as treasures.

Just as we are continually being bombarded by so many messages from advertising and the media that they cease to have any meaning, so, in the products we use, we are often overwhelmed by the excessive number of features they have. The unused and irrelevant possibilities they offer us litter our minds like mental garbage. As a result, we no longer really relate to products or develop any affection for them. We never really make friends with them: they remain just casual acquaintances or incidental business contacts at the edge of our real existence. And, because we do not feel bad about throwing away products we have not become attached to, they end up cluttering, and eventually polluting, the environment.

While we've been creating this quantity-mad world, an essential human need has been neglected. Consider the sort of relationship our ancestors had with their objects – trusty tools, family heirlooms, totems or magical objects. Such objects not only served a practical function. They were also carriers of memories, of personal or family history; they were magical or ritual objects that gave protection or expressed belonging. Because these objects remained in society for a long period of time, they provided a sense of permanence, stability and continuity. In our throw-away society, we've lost this affectionate relationship with most of our contemporary objects. We no longer form an emotional bond with them.

Today, we may cherish a book that belonged to our grandmother – not so much because of its contents, but because, for us, it is somehow part of her. But which of the products we use today will our grandchildren cherish when *we* are gone? To answer this question, we need to become aware of what it is about certain products that makes people

  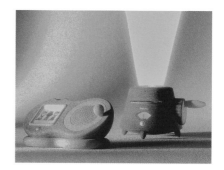

want to possess them even when, in terms of their functioning, they've become outdated. What is it, for example, about classic cars, about veneered 1930s radios, or fifties' chrome; or, going further back, about antiques in general? They have an added value in terms of the cultural aura they emit, a value which in due course may totally replace the original functional value. Our challenge is to look ahead to see what sort of products will form the heirlooms of the next generations – the New Heritage.

**Rent Line** Whereas New Heritage products answer to a maximum degree people's desire to possess and keep beautiful products as carriers of cultural and personal values, Rent Line answers another basic need: to solve a practical problem right now.

We've seen that there are some products we like to create an emotional bond with, to have them with us as our long-term companions. But there are others which we need only for a short time in order to accomplish a particular task: their major desired characteristic is effectiveness. Today, however, we're very often obliged to buy such temporary products on the same basis and terms as those which we plan to keep with us forever. We need to find a new way of thinking about such products, one which takes account of this variable need for possession.

One of the most recent trends in manufacturing and consumption is the replacement of physical products with software and services. One only has to think of the possibilities that have already been realised since the invention of the Compact Disc. The converging technologies of CD, computer, video, telecommunications in the information super-highway are rapidly creating vast new markets for multimedia services. However, a major problem for consumers (and for manufacturers) is that things are moving at such a tremendous pace that anyone who wants to keep up has to continually invest in new hardware. But there's a limit to how much and how often people are prepared to pay for hardware which seems to have obsolescence built into it right from the start.

How can we overcome this problem? We need to ask ourselves what people really want from the service-carrying hardware. Clearly, what they want is the service – the latest service. They have no other interest in the hardware. In other words, the object itself is only a tool, a means to an end. Taking this need seriously will mean providing products on a temporary basis; in other words, renting products out. It will also mean creating a line of products specifically for the rental market. Such products will contain state-of-the art technology but have a simple, inexpensive exterior. Customers therefore pay only for what they want: maximum effectiveness and maximum efficiency. The extra development and production costs involved in creating an aesthetically appealing exterior would be saved. Then, whenever customers decide that the product no longer provides the level of effectiveness they need, they can exchange it for a newer model.

But what does the manufacturer do with all that returned equipment? The answer is simple. It's rented out to other customers whose requirements are less demanding, either in another segment of the same market, or in a less mature market somewhere else in the world. And, of course, if it's been properly designed, it can always be upgraded or recycled it to make newer models.

In addition, as the products move around the world from one user to another, carrying with them, like labels on old suitcases, the signs of their travels, people may find it an attractive idea that a product was once used by someone else in another part of the world. They form the latest link in a global network of fellow-users.

The Rent Line strategy also has important commercial benefits for manufacturers. It sidesteps customer resistance

New Heritage concepts include (left) Affection Stones,
personalised remote controls to operate an ultra-thin
screen, and (centre) Memory Messenger, a leather-
bound device to store sights, smells and
tactile sensations to pass on to our children.
Improve Now! concepts include (right) a storyteller and
a bedside projector that are powered by hand.

to making what feels like a long-term investment in products that turn out to have only short-term usefulness. It also allows them to exploit the different levels of maturity of different markets.

There's yet another human benefit of this strategy. It can enhance the desirability of products such as those that result from the New Heritage strategy. Thus, the television of the future might consist of an attractive New Heritage set, with, hidden somewhere inside, a rented satellite decoder allowing reception of many more services than were available when the set was originally bought. In that way, the consumer gets the best of both worlds: the comforts of permanence together with the advantages of change.

Rent Line products would also have very clear benefits as far as ecology is concerned. The fact that the life of the hardware is extended by being used longer is the prime and most obvious advantage. The greater the number of rented products that are used in different markets, the fewer obsolete products are being discarded.

In addition, upgrading Rent Line products will simply be a question of modifying the service software; the hardware will remain the same. And, of course, recycling software is more environmentally friendly than recycling hardware could ever be. But more important than both of these benefits is the fact that the services accessible through Rent Line products provide information and services which directly result in reduced consumption of matter and energy, by cutting down on the need for transporting people or objects.

**Improve Now!** Strategies like New Heritage and Rent Line are naturally long-term projects. But if they, or something similar, are to be realisable in the future, we need to develop enabling strategies – ways in which products in general can be made more environmentally-friendly. Some of these strategies are already the subject of intense work. They include developing clean technology for the manufacturing process, working out ways of evaluating environmental impact, and so on. Another important strategy is design for disassembly and recycling – looking at ways in which products or parts of products can be re-used at the end of their useful life. Three other strategies that I'll deal with here are durability, miniaturisation, and usability.

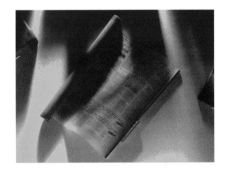

Rent line concepts include a digital camera, a digital
newspaper and an electronic map

Durability   Both New Heritage and Rent Line depend crucially on the possibility of creating durable objects. Durability can be achieved in two ways. One is to make objects which, though only moderately durable, are nonetheless highly recyclable. The other is to extend the life-cycle of products; that is, to develop a generation of products which will work for a very long time. The practicability of the first strategy depends very much on the availability of solutions which make the repeated use of materials and components economically viable: the actual construction of the products, systems of collection, the amount of energy needed for recycling, and so on. The practicability of the second strategy depends on the development of materials and surface treatments which can be maintained by the user and which 'age well'. They must decay as slowly as possible and through a series of stages which in cultural terms can be seen as an improvement in quality. Like wood, for instance, or leather or copper, the older they are, the more attractive they become. This is, of course, crucial for the New Heritage products.

We tend to think of durability as applying only to products. But it could apply to the packaging, too. Packaging could be made to be directly biodegradable, or highly recyclable; or, in certain cases, it could – in line with the New Heritage idea – itself be made more durable in the form of, for example, an attractive container, which can be kept longer than the object it once contained.

Miniaturisation   The strategy of miniaturisation involves making objects of minimal environmental impact by cutting down the amount of material and energy involved and generally making them smaller. The practicability of this strategy depends on the availability of high-performance materials and technologies to improve the performance-size ratio of products.

The Rent Line strategy will often involve moving the goods from one part of the world to another. The smaller the products are, then, and the lower the storage and transport costs, the more the idea becomes economically feasible. For New Heritage, a particularly interesting development resulting from miniaturisation is that products are increasingly nestling into our bodies. The size of Walkman earphones is an obvious example. They're becoming smaller all the time. As this process continues, products are becoming more like jewellery. How long will it be before earphones are indistinguishable from earrings? At that point, we move into the New Heritage field. Will our high-end decorative walkman earphones be passed down the generations to be worn as earrings by our grandchildren? Will the microphone of our portable phone be incorporated in a ring, which will be treasured long after we've made our last call?

Usability   The strategy of usability, naturally enough, focuses on creating products which users find very easy, efficient and pleasurable to use. Usability in terms of maximal clarity and comfort is obviously important for Rent Line products. And since New Heritage type products are designed to generate affection in the user, a vital part of their success will depend on how pleasurable they are to use. Notice that, if miniaturisation continues at its present rate, it will inevitably change the channel through which that user-product interaction takes place. Something which is comfortable to use when it's large may need to be operated in a completely different way when it's small. As computers become smaller, for example, operation by means of a keyboard is becoming less and less feasible – at a certain point, even the most delicate fingers become too clumsy and inaccurate. Consequently, voice activation will become inevitable, and we shall converse with our machines as if talking to a friend – or to ourselves.

**Relevant Functionality**   I just mentioned the sense of alienation that develops when a product has so many features that the user becomes confused or has no need of them. The New Heritage strategy, by endowing the product with cultural and personal values, is one way of countering this reaction. Another way is to ensure that the functions that a product offers the user are precisely those that are relevant to the user's needs.

Take, for example, the telephone. Many models these days try to incorporate all the functions that are technically realisable: not only making a simple call, but passing calls on, muting, call-back function, diverting to another extension, storing numbers in memory, and so on. Yet most people only want to make and receive calls. Nevertheless, even to do that, they have to find their way through the complex pad of keys and buttons. Although they do not use such functions, these people have paid for them; and valuable resources – of time, money and human effort – have been devoted to implementing those functions. In both respects, a multi-functional phone for those particular users constitutes a substantial waste.

The strategy of relevant functionality would require that consumers could basically customise their own products, by selecting the functions they require at the moment of purchase. Ideally, of course, they would be able to add to these later, as the need arose. For some people, a mono-functional product would be quite sufficient; others might want to incorporate all available functions. As more and more functions come to reside in the software rather than the hardware, or at least in easily replaceable programmable components, such customisation will be easier to implement.

**Not just a question of conscience** Application of strategies such as those I've just described will improve people's lives. But the sales of ecological products cannot depend on consumers feeling morally obliged to buy them. They have to *want* them. Only when people actually like what's good for them are we on the right track. We cannot take the short-term view, looking for eco solutions, finding them and then sitting back on our greener-than-green laurels. Eco labels cannot be relied upon to sell products, because, although 'green products' may sell well today, eco fatigue may set in tomorrow. To achieve its purpose, Eco Design must create Eco Charisma.

 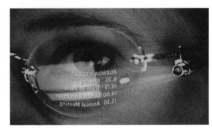 

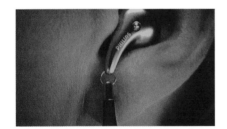 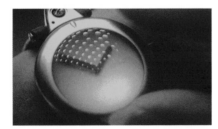 

Miniaturisation means that communications media can be incorporated into jewellery and fashion accessories, such as tiny ear-ring earphones, screen-phone watches and glasses that project information onto the lens (concepts from *Vision of the Future*)

# Talk to me, Moses!

Let me transport you for a moment to the Eternal City – Rome. And let me take you from the bright Italian sunlight of the street into the dark, cool interior of the Church of San Pietro in Vincoli. There, after a little searching, we come across the tomb of Pope Julius the Second, and in particular, a magnificent statue of Moses, one of Michelangelo's finest sculptural achievements. The story is told that, when he'd put the final, finishing touches to the statue, the maestro stood back to examine his work. It was so life-like. "What do you think?" he asked. But he wasn't talking to his apprentices, who were all standing around admiring the result of months of work. He was talking to Moses. But Moses kept his silence. Then Michelangelo, to the horror of everyone looking on, strode forward and gave Moses a mighty blow on the knee. "Talk to me, Moses, talk to me, damn you!" he shouted angrily. "I've brought you this far – the very *least* you can do now is say something!"

**A question of behaviour**  Apocryphal or not, I think this story shows us just how far we have come in the past four hundred years. Not that we've surpassed or even equalled Michelangelo's artistic genius, but we *have* reached the stage now when we can make our creations talk back. We now possess the technologies to make our products, the objects we create, respond to their users. And increasingly, these objects are no longer just reacting to our actions; they are becoming subjects, independent actors on the stage of human life, interacting with us and sometimes even taking the initiative.

The question that now needs to be tackled – and the one that Michelangelo was spared – is: "What do we want Moses to say?" How do we want our interactive and even proactive products to behave? The age of *Star-Trek*-style humanoid servants is no doubt still far ahead of us. But an age in which devices are independent enough to undertake actions on our behalf is already here, and if we're to avoid a confusing and even dangerous chaos, we need to lay down the ground rules for the behaviour of such devices.

This is a serious obligation. As designers involved in designing interactive systems, we're designing devices that, within the next few decades, will have invaded almost every corner of people's homes and every aspect of their lives. This imposes a great ethical responsibility on us, because a product that seems to offer a clear benefit on the micro-level may turn out to have catastrophic consequences when implemented on a grand scale. Lyle D. Goodhue, the inventor of the aerosol, was hailed as a hero back in 1941, when he came up with his bright idea. Now, fifty years later, we know what aerosols (or rather, to be fair, their CFC propellant gases) can do. If we'd known then what we know now, we would *not* have had a hole in the ozone layer and a potential ecological disaster hanging (quite literally) over our heads.

**Balance**  So designers of even the most humble, everyday objects have a great responsibility – a frightening responsibility – to look ahead and consider, at the outset, all the possible implications their product may have in the future so that they can try to avoid any negative effects. This is all the more vital, because one of the main motivations inspiring designers in their work (besides the need to earn a living and a thirst for intellectual stimulation) is the desire, through their work, to improve the quality of people's lives. And quality of life implies balance: balance between people and their natural environment; balance between people and their artificial environment; and balance among people themselves. This is ultimately what is meant by the sustainable society: a society which values and cares for its own extended eco-system so that it can continue to exist indefinitely.

**Social and economic balance**   The realisation that this holistic balance is necessary has come about only gradually. It was the social consequences and the excesses of the Industrial Revolution that first aroused people's awareness of the need for more social and economic balance within society.

And, for our part, as designers and manufacturers, we've contributed to establishing this balance by spreading the material comforts of modern society more widely, essentially making it possible for everyone to experience benefits which used to be limited to the rich and powerful – the benefits of having servants, the benefits of being well informed and entertained, the benefits of being mobile, and so on.

**Environmental balance**   More recently, the environmental shocks of acid rain, oil spills, Chernobyl and now the ozone layer have driven home the fact that we're all responsible for looking after our natural environment. We need to recover the balance here, too. So, as designers, we now make sure our products are as environmentally friendly as possible. We reduce their impact on the environment by making them highly recyclable, or highly durable; we make sure their entire life-cycle is clean. And thanks to miniaturisation, we can cut down on the amount of material and energy needed to make, deliver and use those products. Software is replacing physical products or parts of products so that they can be easily upgraded and don't have to be entirely replaced. And by linking information technology with telecommunications, we're creating further environmentally beneficial developments, such as teleworking, telebanking, and so on. So I don't think we can be accused of not doing our best on this score. In fact, by promoting greater social equality, encouraging a more even spread of economic wealth, and fostering 'green' products and behaviour, society is essentially taking out an insurance policy on behalf of future generations. All these developments are precautions that have a cost. And that cost is the premium we pay now, so that our descendants will not be confronted with violent upheavals or ecological disasters.

**Artificial environment**   But, of course, the social and the natural environments are not the only environments we inhabit. And they're not the only ones we need to insure for the future. We are also surrounded by an artificial environment, an environment of our own making, full of products; an environment for which we as designers bear a special responsibility. And just as our social structures can be threatened by glaring inequalities and our natural environment can be polluted by pursuit of the quick fix rather than the long-term solution, in the same way our artificial environment can be polluted by products that create as many problems as they solve.

In fact, we humans are already in danger of being crowded out of our own home territory by the things we have brought into it. I don't mean physically crowded out, of course, because miniaturisation is allowing us to create *more* space for ourselves. I mean we're being mentally crowded out – overwhelmed. We're suffering from information overload: advertising shouting to us at every street-corner; junk mail blocking up the letter-box; the World Wide Web tempting us to let the waves of useful (and useless!) information wash over us, till we're thrown back, exhausted, onto the shores of the mighty Cybersea (either that, or until we get worried about the phone bill). In the end, and at its worst, it results in a severe case of mental pollution. The same goes for all those products that have too many features, which just end up confusing the user, who probably never needs them, anyway.

As designers, our response to this sort of pollution in the artificial environment has been to develop techniques of enhancing user-friendliness. To strike a balance between what is technologically possible and what people actually want to do – and can comfortably cope with. But simply preventing things from getting worse is not really what we're after, of course. If we, designers, want to improve the quality of life, we also have to make sure that by improving things in one respect, we don't end up making things worse in another: remember the aerosol. But how can we do this?

**Friends of the user**   One way, I suggest, is to make sure that products are not just friendly towards their users but that, in a much more comprehensive sense, they become friends of the user. After all, we all like to be surrounded by our friends, the people who make us feel good, who help us achieve our goals: all-round enhancers of our lives, in fact. So if we can make products which do the same for their users as our friends do for us, we'll be improving the quality of people's lives significantly. As designers, we need to find ways of turning dead objects that leave us cold into subjects with a spark of life, things we can truly warm to.

**From object to subject**   That's not as difficult as it sounds – at least, as long as we don't try to do everything at once. We can break it down into manageable chunks if we recognise that the dividing-line between objects and subjects is a fuzzy one. It's the result of the interaction of many different characteristics, independent parameters along which products can be ranged, from archetypal 'dead things' at one end, to products that are experienced as almost 'alive' at the other. The result is more of a continuum than a categorical split.

It's somewhat like the distinction between strangers and friends. You can't just classify everyone as belonging to either the one category or the other. Life's more subtle than that. There are complete strangers who do not affect your life in any way – you may pass them in the street, but you hardly see them. And there are other strangers who very clearly *do* affect your life – ranging from the paramedic who helps you into the ambulance after an accident, or the sales assistant who helps you choose a tie, right through to someone who becomes your future lover. The same goes for friends – it's a broad category, ranging from colleagues and neighbours, to old college friends, soul mates, and partners.

What we need to do is find ways of moving our products along that continuum, away from the object end, and towards the subject end. To make our task easier, we need to isolate appropriate parameters, to find the proper characteristics, which we can then manipulate to bring about this shift from object to subject. Let's look at just three of these characteristics or parameters as a start.

**Relevant products**   The first parameter is relevance. To be wanted, products have to be relevant, that is, useful to their user. The usefulness of a product depends very heavily on its context: the identity of the user, the time, the place and the circumstances. Take a lawnmower, for instance. If you have a large garden, it could be very relevant to you, in terms of helping you keep the grass tidy. But if you live on the top floor of a high-rise apartment building, it's probably the last thing you want cluttering up your living space.

**Meaningful products**   The second parameter is meaningfulness. To be loved, products have to be meaningful to their user. Meaningfulness and relevance are independent of each other. Take the lawnmower I just mentioned. It could be meaningful – or not – to either the garden-owner or the flat-dweller.

Suppose the garden-owner didn't care which lawn-mower he used to cut his grass. The machine would be relevant, but not meaningful to him. But suppose he was very attached to his particular machine – perhaps because it was so light and cut so well, or because the sound of it reminded him of warm summer evenings in his childhood, and his father cutting the grass. That would make it both relevant and meaningful. For the flat-dweller, we said, the lawnmower would be low on relevance. But it could still be very high on meaningfulness. Suppose he was once a professional gardener, and was now living in a retirement flat. It might represent for him the last memory of happier times. In other words, meaningful products are those that become something *special* to the user, something the user cherishes for non-rational reasons.

Suppose, as designers, we deliberately try to incorporate as much meaningfulness into our products as possible. We may not be able to attach specific memories to them, but we can make them so that it is easy for people to absorb them into their lives. We might do that, for instance, by making a kettle that isn't just a device for boiling water but is an attractive part of the kitchen environment, enhancing the social ritual of making tea, with its pleasant associations of entertaining friends, taking a break, having breakfast with the family, and so on. This sort of product combines relevance with meaningfulness. And like the faithful butler, perhaps, it is well on the way to becoming a friend of the family.

That's more or less what we at Philips were trying to do with our Philips-Alessi Line of kitchen appliances. To make relatively humble objects into something special, something more meaningful. Because when a product has meaning for its user, it not only gives pleasure; it's less likely to be thrown on the scrap heap at the first sign of ageing. And that's something that will ultimately benefit the environment. In fact, the more we succeed in giving products that added value of meaning, the more we shall be contributing to the attainment of a sustainable society. And by making products more meaningful, of course, we're essentially making them more human. Human beings are much more interested in other human beings than they are in dead objects, so the more we're able to design objects that trigger lasting interest and affection in their user, the more we will be humanising them. Technology is increasingly giving us the chance to take this process further still. But we have to make sure we're giving our products the characteristics we like to experience in our friends. So what is it in others that attracts us?

**Personality**   I suggest that the secret ingredient is personality – and this is the third parameter. It's a word that has a number of different meanings, but we could define it, along with Webster, as "the totality of an individual's tendencies to act or behave, especially self-consciously; acting on, interacting with, perceiving, reacting to, or otherwise meaningfully influencing or experiencing their environment." In other words, it's the way people behave towards others; and the things they do – or don't do. Now, products are becoming increasingly capable of acting. Able to interact with us, and soon to 'proact' as well (if that's a verb), capable of being proactive, of taking the initiative in interactions, rather than just reacting to something the user does. As our products move further along the continuum from object to subject, they're adding a new dimension to our artificial environment: the dimension of independent (or quasi-independent) behaviour. And this is a dimension that the designers of interactive systems in particular must design carefully if problems are to be avoided in the future. We need to be able to design the sorts of personalities we want our new companions to have.

**'Civil commotion and riots'**   When discussing the natural environment above, I mentioned insurance policies, and the fact that we as designers have a duty to think ahead and build them into our creations. The same applies to the artificial environment. The small print in an average home insurance policy shows that the holder is not only insured against the usual disasters, like fire, lightning and explosions, but also against damage caused by "strikes and labour disturbances" and, even more significantly here, "civil commotion and riots." That's precisely what we'll have to insure our descendants against in the near future – "civil commotion and riots" – as our products, no longer just objects but subjects with minds of their own, start doing things for us: each of them with the best of 'intentions' no doubt; each, on its own, providing a significant benefit; but together producing a chaos that drives you crazy.

**Ground rules**   I've recently been struggling to adapt to some new word processing software. Unlike my previous program, it does a lot of things on its own, without my asking it to. Which is nice – some-

times! But sometimes it does things I wish it wouldn't do; and I still haven't managed to work out how to stop it doing them. Not quite commotion and riots, perhaps, but nonetheless very annoying. It's nice having things done for you, but it's nice to feel you're still in charge. Even when you're not.

So, as designers, we're going to have to develop patterns of behaviour for these new 'objects-almost-subjects'. We're going to have to lay down the ground rules which govern relations, not only between them and us, but also among themselves. Consider the case of a smart home in which various products are activated as we arrive home from work: the message device reports on who's called, the audio system plays our favourite music, the cooking centre tells us how dinner's coming on, and – Oh, it's six o'clock! Time for the news! – so the TV flicks on automatically as well. If you've had a hard day at the office or a tiring journey, the last thing you want is to be greeted with a cacophony of sound as all these proactive devices compete for your attention.

Or take the well-known problem of mobile phone etiquette. Wouldn't it be useful if the phones themselves just *knew* when it was appropriate to ring, or when it was appropriate to give us a metaphorical tap on the arm, or when to put off giving us the message until later?

These are simple examples, perhaps, but they illustrate my point well enough. We need to educate our products; or rather, we need to build good, social behaviour into them. After all, in a very real sense, such products are the modern-day equivalent of trusted servants. And of course, in the old days, if you were fortunate enough to have servants, you saw to it they were trained to perform their job discreetly and unobtrusively. They were taught to act appropriately, depending on factors such as the time, the place and the circumstances.

Like new servants, like new members of staff in an organisation, or, indeed, like new-born babies, our new products need to be educated in the morality and culture of the society they're entering. They need to be socialised. Their personalities need to be shaped. We ourselves have been through all this. Now, as inanimate objects grow up to become independent entities, they need to undergo the same process. Of course, the difference between these new quasi-subjects and ourselves is that we were allowed a couple of decades to achieve an acceptable level of socialisation. We shan't be so patient with machines. We'll want them to emerge fully socialised from the 'womb', as it were. They'll need to be plug-and-play, ready-to-use, grown-up and sociable, right from Day One. So what we as designers have to do is not simply design the objects, we have to design their subjective natures and their social behaviour as well.

**Design government**   I like to think of the designer's role here in terms of the metaphor of government. When the Ancient Greeks and Romans established the basic form of our present democracies, they laid down rules to govern the relationships between members of the community so that a balance of interests might be maintained. In the same way, we now have to lay down the rules of the new community, in which humans will be joined by a whole sub-community of proactive devices. In terms of this governmental metaphor, designers are the members of the senate or members of parliament. They represent the people – the consumers. In drafting the legislation (i.e., the rules governing product behaviour) the designers work together in cross-party committees (i.e., multi-skilled teams) to make sure that the interests of a broad cross-section of the people are represented.

**Multiple skills**   Let's deal with these teams first. They have to be multi-skilled, because what we're trying to do is so complex. Understanding what people really value in a product is such a complex matter these days because people, their needs and the products themselves are so multifarious. It goes beyond the capacities of any single individual. Whatever our own discipline, we need to look outside it and see how others can help us refine our work still further. In other words, we need to understand much more about human psychology and motivation. Technological expertise and insight are not enough. They need to be balanced by an equal level of 'people expertise'. The members of parliament, as it were, need to get in touch with their grass roots, the people they represent – the men,

women and children in the street. Each of those people, potential users of products, is a multidimensional human being. By studying people with respect to all their characteristics and dimensions, we should be able to create products that they experience as not only relevant, easy to use, and attractive, but also as products with personality.

At Philips Design we are tackling this problem of discovering what product qualities are valued most highly in a number of ways. Besides specific research on trends in demographics, lifestyles, psychology and social anthropology, we mount exhibitions in which we try to visualise in a very concrete way various future possibilities. The aim is to trigger reactions from consumers. In this way, we're able to draw certain conclusions about the sort of things people like and dislike. This feedback on realistic models is more revealing than simply asking people what they would like, because until they can see or feel something, they're often unable to be specific.

**Documentation**   It's important that designers carefully record their findings in the field of 'people studies', so that they're subsequently able to repeat their successes and avoid repeating their mistakes. In a sense, in our search for knowledge about consumers, and in developing skills to produce products and services they want and need, we're entering a period of explosive scientific data-gathering not unlike the great nineteenth-century boom in the natural sciences. We need to develop ways of registering and classifying our findings, of imposing order and under-standing on them, so that we can expand and refine them, without retracing our steps or duplicating the work of oth-ers. Whenever we at Philips mount one of our exhibitions, for example, we keep a comprehensive record of the prelim-inary research, the feedback obtained and the methods we used, and some of this material is published in the form of a book or a Web site. Even in our normal product creation activities, we follow a clear process model, documenting work as we go, so that we can spread the knowledge and insights gained along the way to all parts of our global service unit.

**Managing the balance**   Finally, in our model of a design government, I haven't men-tioned the government itself. Someone has to manage the balance within the artificial environment, lay down policy guidelines and give direction to this work. And surely, as designers of interactive systems, we're the ones who have to take the lead here. We, after all, design the behaviour of those objects-turned-subjects, the products that will soon 'peo-ple' our artificial environment. These products will need to behave like well-brought-up, responsible adults, people we like to have around. It will be our task to plan the education, the upbringing, the socialisation of this new generation of products. It will be our job to make sure tomorrow's artificial environment is pollution-free, a world of personalities and friends, a world we can be proud to have created. The premium we pay now in terms of extra care and foresight will have been well worth it if we can ensure that our descendants will live in harmony with the results of our efforts.

**Beyond Moses...**   And what about Moses? What would he have said to Michelangelo – if he could have talked? More importantly, what will *our* products say to us in the future? Are we going to let them embar-rass us in public? Or are we going to make sure they're well-integrated members of the artificial environment? Providing we educate them properly, if we follow our user-inspired commandments to make sure they're relevant and meaningful products with attractive personalities, then I'm confident we'll have nothing to be ashamed of. On the contrary: we'll be able to take credit for improving the quality of people's lives by providing them with products that are real friends.

Michelangelo was frustrated by the fact that his Moses would not talk. We can now make our creations talk: *our* challenge is now to do more. We have to make sure that our creations also behave responsibly. If we do that, we shall be completing the work of one of the greatest artists of all time. Surely, no mean achievement.

# The Monk and the Machine Gun

The front page of an Italian newspaper recently featured a fuzzy black-and-white security camera photograph of a bank raid. It showed a monk, with his hood over his face, wielding a submachine gun at the startled staff and customers. The man escaped with a substantial sum. But who was this man? A man of God who had taken a vow of poverty and had mentally flipped, suffering, perhaps, from some sort of Robin Hood complex? Or was it a violent criminal with earthly motives who had donned the habit of a monk to lull his victims into a false sense of security when they saw him enter and to conceal his weapon from them until it was needed? We know enough about monks and violent criminals to make a good guess as to the answer – after the event. But this incident neatly illustrates some important principles of branding. First, it shows how people rely on their expectations, based on past experience, to guide their response when they see a familiar symbol. It shows, too, how the use of a given symbol does not necessarily guarantee that the qualities which the symbol represents are really present: in other words, symbols can be misused. And thirdly, it shows how important it is, therefore, to safeguard the integrity of one's brand.

**Branding – the urge to be distinct**   So much is talked about the importance of branding these days, and how it can best be achieved, that we tend to lose sight of its essence – its fundamental purpose and significance. It may therefore be useful to go back and re-examine its early beginnings. Branding was originally a way of identifying one's livestock or other goods by burning a mark in them with a hot iron, distinguishing them from those of others. It was in a sense a reflection of a primitive need to individualise, to present oneself as distinct from others. And such marks of distinction were not limited to individuals, but were also used by larger groups in society and even whole nations. Military uniforms, insignia, standards, coats of arms and flags all served the purpose of expressing and strengthening the homogeneity of the group. Through them, soldiers gained a greater sense of coherence as individuals sharing a common cause. In practical terms, the 'branding' signs also helped them to identify each other in the thick of battle, and led their enemies to perceive them as more unified and therefore more formidable than would otherwise have been the case.

Later, in the commercial sphere, the brand – now not only burned marks but also other marks, names or symbols (trade-marks) – became a way of indicating not so much ownership as origin (maker or supplier), and therefore also a certain quality distinct from that offered by competitors.

**Expressing abstract qualities in visual form**   At a certain point, a simple statement of makership (such as 'I, Hlewgastir, son of Holti, made this horn') was not felt to convey distinctly enough the special characteristics implied by the maker's reputation. How could such abstract concepts best be represented?

The way in which the maker's name is written can, like handwriting, convey more than just the name itself.

The logotype – the stylised form of a brand name – is the corporate equivalent of the artist's signature. It exploits the associations that certain typefaces have. Some of these associations are inherent, in that they are largely universal. For instance, rounded forms tend to imply 'smoothness', 'femininity', and so on, whereas more angular or more solid forms imply 'toughness' and 'masculinity'. Similarly, the use of slanting letters to convey 'speed' is based on their ironic resemblance to the angle of the human body when running.

Other associations may be culturally determined. For example, in a culture with a strong tradition of printing, typeface that resembles handwriting has an air of informality. The use of the Roman script for imperial inscriptions gave it associations of authority, and later also of conservatism. The angular Fraktur script formerly used in German-speaking countries also developed strong cultural associations. Deriving from its use in Martin Luther's translation of the Bible, it became a symbol of national pride following German unification in the 1860s and was later adopted as a graphic symbol by the Nazi movement. Today, its attributes in Germany are 'ornamental', 'old', 'traditional' and 'conservative', although in certain contexts it can still evoke the aura of the Third Reich and through that association also be perceived as threatening.

Another way of conveying abstract qualities is to use an index or symbol. Many trades, for instance, used as their mark the image of one of their tools (e.g., a pair of scissors for tailors) or of the goods they supply (e.g., a loaf for bakers). Some countries, similarly, have chosen images of their local flora and fauna (e.g., the maple leaf for Canada, the shamrock for Ireland, and the kangaroo for Australia) as their identifying visual element. Other nations have chosen a symbol of the power they wish to be seen as possessing; thus the Romans chose the eagle, while the British chose the lion. A modern corporate example of the use of a symbol is the Legal and General insurance company's use of an umbrella to imply protection against adversity. Such attempts to solve the problem of how to express in visual form something which is abstract all have some relatively easily traceable link to an object in the real world.

What about shapes that have never had any clear link to the real world, such as geometrical shapes? No doubt forms with angles are perceived as harder than, say, circles or spheres; and the circle has also traditionally been interpreted as a symbol of perfection. But these forms, too, can acquire cultural meanings. Thus, the diamond may be perceived as having associations of gambling, if the link with playing cards is made.

Finally, colours can also be used to convey abstract qualities. The associations of some colours can be relatively easily traced: thus, red is often taken to imply aggression and anger (through the fact that a flushed complexion is often indicative of anger and aggression). Bright blues, greens and yellows, being the predominant colours of spring in non-tropical latitudes (the season itself symbolising a new beginning), are often associated with cleanness and freshness and therefore used in the marketing of household cleansing products. Other colours have associations that are culturally determined: the use of pastel colours to imply femininity and dark colours to imply masculinity would seem to be based on a correlation between colour saturation and the culturally perceived relative 'strength' of the sexes. A change in these cultural values may eventually lead to a change in the associations of such colours.

**Cognitive dissonance**   In all these sorts of cases, individuals and groups seek to convey something of their character and qualities through a visual representation. But visual appearance is not enough to convince people that the particular qualities they promise can indeed be delivered. There must be a coherence between appearance and reality – people must act in ways that fit in with their image, companies must provide the quality of goods that they appear to be claiming in their communications, and political parties and countries must behave in ways that fit in with the claims contained in their names (e.g., 'democratic republics' should be democratic). If this is not the case, then the result in the mind of the observer is what is known as 'cognitive dissonance' – a perceived mismatch between what is claimed and what is provided, between what is said and what is done.

**Communication is time-bound**   But although a clear visual representation of the qualities or characteristics one stands for and some clear behavioural evidence that what is claimed and what is delivered are in agreement with each other are necessary conditions for the success of a brand (an individual, group or company, etc.), they are not sufficient. What is required in addition is that whatever is claimed or offered be appropriate to the time and culture in which it finds itself. The prophet crying in the wilderness, the political idea that is ahead of its time – no matter how true to their image they may be, they do not succeed, because the society in which they appear is not ready to receive them. Hilter's fascism was unsuccessful to start with because German society was not ready for it until the Depression and the political stalemate in the Weimar republic around 1931 made it seem a viable alternative. Similarly, in recent British politics, centrists in the Labour Party went from being outcasts in the late 1970s to being the successful heart of the party in the 1990s. The intervening experience of the Thatcher 'revolution' and the general collapse of Communist regimes around the world had established the non-viability of extreme leftwing politics, at least for the time being, and made the majority of the electorate ripe for a more centrist party.

**Changing dreams**   The world of politics is only one of the more obvious areas of activity in which such swings of opinion and taste result in dramatic success or dramatic failure. But the same thing is found in industry and commerce, and even in personal relationships. Yesterday's philanderer becomes today's sexual harasser, and yesterday's well-brought-up young lady becomes today's spoilsport, for instance.

Consider the history of brands over the last fifty years in the United States. After World War II, they symbolised the good life: solid values, happiness – the attributes of a life worth working for; in short, 'the American dream'. As economic prosperity grew, they came to symbolise in the sixties and early seventies 'share the wealth' – everyone can have the best – let's enjoy together. Later, as the oil crisis struck and economic problems arose, confidence waned. The previous message no longer worked, because clearly not everyone could afford the more expensive branded articles; and the traditional image of cars (the bigger, faster and shinier the better) was taking a beating as it became necessary to reduce fuel consumption. Then, in the competitive early and mid eighties, the branded article became a symbol of superiority: a BMW; a Lacoste shirt, Gucci shoes and a Rolex watch were all seen as attributes of personal success. In the late eighties and early nineties, following the stock market crash and during the subsequent recession, brands came under threat from private label products, as people began to resent paying a premium for products with a name, when they could get something roughly equivalent cheaper by shopping around. From the mid-nineties, however, partly because of the accelerating pace of life and the increase in the range of choice, people have started returning to brands as reliable beacons in the 'chaos' around them.

**Nothing is forever**   It's not only social developments that can change the value given to brands. The technological context, geographical location, or type of activity being pursued can all lead to differing expectations and interpretations of brand qualities. Take 'fast', for instance, a term that many companies apply to their services, even elevating speed to a characteristic of their brand. When speaking of today's mail service as 'fast', we have, in absolute terms, a different meaning in mind than when using the word in connection with the pony express; and if electronic mail becomes the normal mode for the transmission of written messages (as we may reasonably assume), then within the next few years, the concept of speed in relation to this context will have changed its absolute meaning again.

Similarly, within a single time period, 'fast' will designate different speeds depending on the location and field of activity. A regular traveller to New York who is not a US-resident, for instance, may speak of getting through customs at JFK fast, and mean a speed that an EU-resident more used to Amsterdam Airport may experience as frustratingly slow. And a book written fast anywhere is always going to be written slower than a meal eaten fast is eaten. What's important here is clearly the *relative* meaning of the word – i.e., 'at a greater speed than expected in the context'. It is this meaning that's assigned a positive value. A company that wants to successfully promote its services as 'fast' must therefore be fully aware of the meaning of this term in the relevant contexts of time, place and activity. The example of speed is perhaps one of the simplest we could have chosen. Many other qualities which companies may wish to communicate are much more subtle and complex.

I wouldn't wish to claim that there are no absolute values at all and that everything is relative. But if there are absolute values, then they may still tend in practice to be interpreted differently in different contexts ('honesty', for instance, is notorious in this respect). Certainly, the way they need to be communicated will vary from time to time. The failure of many established religions to realise this fully enough is one of the main reasons for declining attendance at their services, and, conversely, for the rise of sects and other alternative religious and spiritual movements, which have found new ways of communicating comparable values.

**Understanding the context**   To communicate one's qualities successfully, then, it is essential to understand the social, psychological and cultural framework in which one's message will be received. That framework is a complex amalgam of the hopes, desires and expectations people have, both as individuals and as members of a social group, together with their artificial and natural environment. A company therefore not only needs to offer people products and services that they want at a price that they're prepared to pay for them, but it also needs to communicate what it has to offer in ways that are easily understood by its potential customers within the context of the time and culture in which they find themselves. In other words, it has to speak to them in ways they experience as relevant and meaningful.

**Adaptability**   At any given time, a whole range of trends is in progress in society which affect the degree to which people are receptive to certain messages, and, indeed, certain formulations of messages. Such developments will also affect the type of products and services that satisfy people's needs and desires. Today, for instance, in advanced economies, as their populations age, people are increasinlgy conscious of their state of health. This means that they are also interested in health-oriented products and services.

Organisations need to formulate their goals broadly enough to be able to adapt to changing circumstances as they gradually develop. They need to take account of their particular location in time and space – the community in which they wish to sell their products and the historical moment – and how their product and its mode and place of delivery fit in with that space/time context. They need to ensure that at all times there is a good 'fit' between these elements, that they form a coherent whole.

Consider, for example, the profession of blacksmith. As long as the horse was the standard means of transport and source of traction power in agriculture and industry, the blacksmith was assured of a profitable business. But when the horse's role was taken over by the motor car, the blacksmith needed to adapt or go out of business. Many adapted by applying the essence of their art – bending iron into precise shapes – to a new field, making fences, gates, and grilles, for example. At the same time, carriage-makers adapted their art to that of making the new horseless carriages.

In similar fashion, a company like Philips has moved from making light bulbs to making valves for radios, and radios as a whole, followed by the large glass 'bulbs' required for the 'new' radios – televisions. Its business has continually moved ahead, adapting and expanding step by step as the social, cultural and technological context changed.

It's no use living in the past. Sometimes a company needs to change, becoming a different type of company rather than simply adapting slightly. The world is full of companies which began life in a field in which they are no longer active. But in taking each step, it is important to remain the best-in-class – the most up-to-date, the most expert, the most responsive to consumer needs. Just as in the case of the blacksmith, who continued to be a good craftsman as his business changed from the shoeing of horses into the making of wrought-iron fences, it is important to carry over professional skills and pride into any new situation. Brands therefore need to develop through a continuous process of context evaluation (What is the 'tune of the time'? What is the cultural context within which we operate?), adaptation and consequent re-positioning. The closer the consistency between the products offered and the requirements of the space/time context, the more successful the company will be. In other words, it is important to know and understand what 'makes your customer tick'.

**Image versus identity**   Today, large organisations can't deal with their customers on a one-to-one, face-to-face basis in the way that the artisans of only a few generations ago were able to do. The world has become too complex for that. There is often a vast physical distance – and unfortunately often also a vast psychological distance – between those who develop the products and those who end up using them. Between them lies a whole chain of people, making, distributing, and selling the products.

Not surprisingly, therefore, companies sometimes find that their products are becoming increasingly unattractive to consumers: because they no longer respond to relevant needs and desires. At this point, there is a danger that the solution will be sought in more aggressive advertising, new claims and an attempt to give the product a 'facelift'. Unfortunately, all too often this results in a growing 'credibility gap' between the claims made for the product and the benefits it actually offers. Changing the message fails to change the product; and in this way symptoms rather than causes are tackled, with at most short-term success. The consumer notices the cognitive dissonance, and is eventually disappointed in the product.

This approach is like the assistant in a menswear shop who simply responds to a customer's expression of dislike for a particular suit by offering the same suit in a different colour, without enquiring exactly where the problem lies – for example, the cut, the size, the pattern or the price. A much more effective method, of course, would be to conduct research into the customer's wishes and reactions to the suit. In the same way, companies need to make sure they continue to understand their customers, because only by doing that will they be sure they can maintain the appeal of their products.

Some companies respond to a crisis of this sort by imitating unquestioningly those competitors who seem to be more successful. By doing so, however, they fail to solve any of their fundamental problems. Since they have not learned to understand the changing wishes and requirements of their customers, these companies are doomed always to be reactive rather than proactive, to follow rather than lead, and therefore also to have to be satisfied with having to work harder to achieve the same financial reward as the leader. IBM recently expressed this distinction very succinctly in its advertising slogan, 'Only the past can be cloned, the future has to be invented.' Copiers have never learned how to make new things. Of course, there's nothing wrong with deciding to produce copies, to make it your goal to produce good copies. Providing you don't claim your products are 'the real things', customers may choose to buy your products in the full knowledge that they're copies, having concluded that the benefit of a lower price is worth more to them than having an original. It's when the copier claims to be an innovator that the problem of the credibility gap arises.

This cognitive dissonance is fatal for a brand, because brands require you to be consistent, to be what you claim to be – to exhibit an identity, not simply put forward an image. The value consumers seek is above all honesty. They want to receive what they think they're buying – and even if, sometimes, they allow themselves to be deceived into a harmless dream (such as becoming irresistibly attractive to the opposite sex through the use of a certain fragrance), still they have no wish to be misled into believing that it is a scientifically provable claim when it is not.

**Virtual companies, virtual products**   As we said, cognitive dissonance can arise as a result of the physical distance between the manufacturer of a material product and the consumer of the product. But this situation is now gradually changing. We're seeing the development of 'virtual companies', who can get very 'close' to their consumers. Providers of electronic services, for instance, can operate on a one-to-one basis with their customers, who sit alone at their computers or in small groups around a television. This closeness, however, is only possible because of the fact that the product they are delivering is also 'virtual', i.e., relatively immaterial.

**The role of design**    The qualities that make a brand distinctive – the qualities that characterise the relationship that arises between the brand-owner and the customer through the product – are intangible and therefore difficult to communicate to those who have not actually experienced them. Here is a task for the designer. For the ability of designers is precisely that of creating a visual representation of the intangible and of converting the immaterial into the material. It's this ability that explains the crucial role played by design in branding.

The brand-owner wishes to communicate the qualities of his products or services to potential customers. Given that the relationship with these potential customers is so tenuous, being based purely on an unexperienced intangible quality, it's essential that the 'message' is communicated clearly and consistently. This clarity and consistency needs to apply not only within a single communication channel, but across all channels through which the message might possibly reach the receiver. In other words, it's not enough to ensure that all one's advertising claims are consistent. The images accompanying these claims must lend weight to them; the appearance, functioning and quality of the products must be consistent with the claims; the claims must be supported by the experience and testimony of existing customers among the potential customer's friends, or of consumer testing organisations consulted by the potential customer; and the places where the products are made and sold must reflect the same values as expressed in all the other channels. In other words, it's not only what you say and do that counts, it's also where and how you say and do it, as well as what others say about what you're saying and doing.

The importance of this clarity and consistency across all channels of communication is being increasingly appreciated: this explains the recent enormous surge of interest in house styles, logos, company 'looks', and more generally, in integrated marketing communications. As traditional channels of communications have become crowded almost to gridlock, people have begun to realise how vital to the strength and success of their brand almost *any* opportunity to communicate its qualities can be, and, consequently, how seriously any cognitive dissonance within that new communicative terrain can detract from a brand's overall effectiveness.

# Index